T0269605

Batch and Semi-batch Reactors

Batch and Semi-batch Reactors

Practical Guides in Chemical Engineering

Jonathan Worstell
University of Houston
Department of Chemical Engineering

Worstell and Worstell, Consultants

AMSTERDAM • BOSTON • HEIDELBERG • LONDON
NEW YORK • OXFORD • PARIS • SAN DIEGO
SAN FRANCISCO • SINGAPORE • SYDNEY • TOKYO
Butterworth-Heinemann is an imprint of Elsevier

Butterworth-Heinemann is an imprint of Elsevier
225 Wyman Street, Waltham, MA 02451, USA
The Boulevard, Langford Lane, Kidlington, Oxford OX5 1GB, UK

Notices
Knowledge and best practice in this field are constantly changing. As new research and experience broaden our understanding, changes in research methods, professional practices, or medical treatment may become necessary.

Practitioners and researchers must always rely on their own experience and knowledge in evaluating and using any information, methods, compounds, or experiments described herein. In using such information or methods they should be mindful of their own safety and the safety of others, including parties for whom they have a professional responsibility.

To the fullest extent of the law, neither the Publisher nor the authors, contributors, or editors, assume any liability for any injury and/or damage to persons or property as a matter of products liability, negligence or otherwise, or from any use or operation of any methods, products, instructions, or ideas contained in the material herein.

Library of Congress Cataloging-in-Publication Data
A catalog record for this book is available from the Library of Congress

British Library Cataloguing-in-Publication Data
A catalogue record for this book is available from the British Library

For information on all Butterworth-Heinemann publications
visit our website at http://store.elsevier.com/

This book has been manufactured using Print On Demand technology. Each copy is produced to order and is limited to black ink. The online version of this book will show color figures where appropriate.

ISBN: 978-0-12-801305-2

To our daughters-in-law

Sandra Worstell nee Garza

Shelley Worstell nee Jackson

CONTENTS

CHAPTER 1

Introduction

CHEMICAL REACTIONS AND PROCESSES

Chemical reactions require time to occur, require energy or release energy, and require a mechanism for co-mingling reactants. We fulfill these requirements by designing vessels called chemical reactors. Chemical reactors provide

- residence time or contact time;
- energy exchange in the form of heat exchange;
- agitation.

Each such design is for a particular reaction; hence, the plethora of different reactors in the chemical processing industry (CPI). However, the batch reactor and its close relative, the semi-batch reactor, are the most common chemical reactors in the CPI.

We are all familiar with batch reactors: we use them every day to cook food, heat water for tea, or make coffee in a percolator. We use them in our kitchens because they are simple devices—we charge them with ingredients, heat their contents to a specific temperature, stir their contents frequently to achieve good heat transfer, reduce the heating rate to let the resulting product simmer, then discharge the finished product after meeting a predetermined specification. We may use our cooking pot as a semi-batch reactor, which means we completely charge one ingredient to our pot, then continuously remove a component from the pot, as happens when we reduce the volume in the pot by allowing water to vaporize, or periodically add additional ingredients to it. An additional ingredient may be water so we can maintain a constant volume in the pot, or we may add salt, pepper, or a variety of other ingredients to the pot, depending upon the recipe we are following.

Our kitchen experience teaches us that batch and semi-batch reactors are highly flexible. Their flexibility allows us to adjust product quality by prolonging or stopping a reaction, to change to a different

Batch and Semi-batch Reactors.

product grade or a new product, and to increase reactor capacity via schedule optimization.[1] Typically, we design a continuous process for a narrow set of operating conditions, thereby minimizing equipment and control requirements. We would have to add considerably more equipment and control mechanisms to a continuous process to obtain the flexibility of a batch or semi-batch process.

By "process," we mean what occurs inside the reactor. If the material in the reactor is single phase and homogeneous, then the process is a reaction. Such a reaction can occur in a batch, a semi-batch, or a continuous reactor, depending upon our design. However, if the material in the reactor is multiphase, e.g., gas–liquid or two immiscible liquids, then it is a "process." In other words, conversion of reactant to product involves more than chemical reaction; it involves multiple steps, some of which are physical, such as diffusion across a phase boundary. If diffusion across a phase boundary or diffusion through one of the phases in the reactor is slower than the chemical reaction, then we define the process as "diffusion rate limited." If physical diffusion occurs at a much higher rate than chemical reaction, then we define the process as "reaction rate limited."

We use batch and semi-batch reactors for reactions and processes involving

- multiple grades;
- multiple, sequential additions, as in the pharmaceutical industry;
- variable market demand;
- small markets;
- unusual product specifications;
- subjective product specifications, such as taste in the food industry and fragrance in the perfume industry;
- rapid response to changing market conditions.[1,2]

The simplicity of the batch and semi-batch reactors allows us to respond quickly to evolving market opportunities. As soon as we have successfully determined the reaction or process conditions in a laboratory batch or semi-batch reactor and established the heat transfer requirements of the reaction or process, we can move to pilot plant-sized or commercial-sized equipment. If the heat transfer requirements are met by currently available equipment, then we can quickly produce product for market sampling. Such rapid response is important in today's global economy where most companies are reluctant to test a supplier's sample on speculation that the supplier might offer commercial quantities in the future. Many companies only test samples of commercially available product; many companies do not have the resources or time to respond to a supplier's research and development organization.

BATCH AND SEMI-BATCH REACTORS

Batch and semi-batch reactors are generally large diameter, metallic cylinders to which we weld end caps, which resemble hemispheres. We usually mount batch and semi-batch reactors vertically on a support structure. This contained volume thus provides the residence time required by the chemical reaction. In most cases, we operate batch and semi-batch reactors isothermally, which means we add or remove heat from it. We transfer heat through the wall of a batch or semi-batch reactor by welding a "jacket" to it, thereby forming an annular flow channel next to its wall. The jacket possesses an inlet nozzle through which fluid enters and an outlet nozzle through which fluid exits the annular channel. If we employ steam to heat a batch or semi-batch reactor, we put the inlet nozzle near the top of the jacket and the outlet near the bottom of the jacket. If we heat a batch or semi-batch reactor with tempered water or hot oil, we put the inlet nozzle near the bottom of the jacket and the outlet nozzle near the top of the jacket. If the contained reaction is highly exothermic, then we require a high fluid velocity adjacent to the exterior wall of the reactor. Welding half-pipe to the exterior wall of the batch reactor provides a high velocity flow channel adjacent to the reactor wall. Half-pipe is a pipe cut across its diameter but along its length. Lengths of half-pipe are then welded end-to-end and wrapped around the batch or semi-batch reactor. We then weld the edges of the half-pipe to the reactor's wall. The batch or semi-batch reactor thereby acquires a corduroy appearance. Cooling fluid then flows at high velocity through the flow channel created around the reactor.

We install "coils" of pipe inside a batch or semi-batch reactor if we require additional heating or cooling to control the reaction. We place the inlet and outlet nozzles for such coils through the top of the batch or semi-batch reactor. As many as four sets of coils can be installed in a batch or semi-batch reactor. In many cases, such coils act as baffles. It is best to keep coils completely submerged if the reaction occurs in a liquid; otherwise, deposit accumulates on the exposed portion of the coils. Such deposit can cause quality issues over an extended operating period.

To comingle the contained reactants, we run a shaft through a bearing at the top of the reactor. A fixed or variable speed electrical motor mounted atop the reactor supplies power, through a gearbox, to the shaft. We attach an agitator to the end of the shaft to transfer power from the motor to the contained fluid. The type of agitator depends upon the density and viscosity of the contained fluid and whether the fluid is an emulsion, a slurry, or a gas–liquid mixture. In most cases, we attach either a propeller or a turbine to the end of the shaft. We specify the direction of fluid flow from the propeller or turbine: the agitator pumps the fluid either radially, upward, or downward. The height of the contained liquid generally equals the diameter of the batch reactor; in such cases, one propeller or turbine is sufficient to mix the reactor's contents. However, if the liquid height is greater than the diameter of the batch reactor, then we place additional propellers or turbines along the shaft. We also specify the direction of fluid flow from these additional agitators. We do not want adjacent agitators pumping against each other. If the bottom agitator pumps upward, then its adjacent agitator must pump upward. The same is true if the bottom agitator pumps downward.

We equip batch and semi-batch reactors with temperature control that actuates heating and cooling mechanisms per need. In addition to controlling heating and cooling rates, we control reactant addition rate to semi-batch reactors with temperature. We most often achieve fluid level control via weigh cells, either under the feed vessel or under the reactor. We also use metering pumps to control the fluid level in the reactor. Basically, we do not over-burden a batch or semi-batch reactor or process with instrumentation. Because of their simple construction and simple control schemes, batch and semi-batch reactors are relatively inexpensive to obtain and to install.

Thus, the advantages of batch and semi-batch reactors are

- simple construction;
- low installation cost;
- simple instrumentation;
- flexible operation.[3]

These advantages do not come free: batch and semi-batch processes have higher operating costs than continuous processes. Batch and semi-batch processes require more operating staff, which charge reactant to the reactor, monitor the reaction or process, and, finally, discharge product from the reactor. Not only are there more staff, but they are also more experienced than staff operating a continuous process. Batch and semi-batch process staff require extensive technical training since they are making decisions concerning an unsteady-state process while it is underway. Reducing such a staff requires installation of more expensive, sophisticated instruments that can control an unsteady-state process.[1]

DESIGN OF BATCH AND SEMI-BATCH REACTORS

We base our design of batch and semi-batch reactors on the residence time required to produce specified product. We establish that residence time requirement in laboratory-sized reactors. We generally overdesign commercial-sized batch and semi-batch reactors to allow for unexpected problems and unforeseen opportunities.

Moving from a laboratory-sized batch or semi-batch reactor means the heat transfer surface area to reaction volume ratio decreases. Most reactions or processes in laboratory-sized reactors are reaction rate limited, or, sometimes, diffusion rate limited. However, most commercial-sized batch and semi-batch reactors are heat transfer rate limited. Therefore, we must determine the optimal ratio of heat transfer surface area to reactor volume when designing a batch or semi-batch reactor.

Moving from a laboratory-sized batch or semi-batch reactor entails changing wall effects. Since we perform most laboratory-sized reactions in glassware, we may not be aware that the chosen steel for our commercial-sized reactor catalyzes the reaction or process. In this case, the commercial-sized reactor will be more efficient than our laboratory-sized reactor. The inverse is true if the material of the commercial-sized reactor inhibits the reaction or process. If we are aware of such a catalytic wall effect, then increasing reactor size decelerates the reaction or process because reactor wall surface area does not increase as rapidly as reaction volume.

Corrosion is not an issue in laboratory-sized glassware reactors; it is an issue in commercial-sized steel reactors. Unfortunately, many times this point is overlooked in our haste to produce commercial quantities of a new product. Overlooking the possibility of corrosion can damage or destroy a commercial-sized batch or semi-batch reactor. To avoid this situation, we must conduct long-term corrosion experiments in the laboratory while determining the operating conditions for the reaction or process.[1]

We can begin producing product in commercial-sized equipment with meager, imprecise data if we use a batch or semi-batch reactor. However, once the market will support it, we must optimize the reaction or process. Optimizing any chemical reaction or process requires knowledge of the chemical kinetics underlying it.[1]

SUMMARY

This chapter introduced chemical reactors in general and batch and semi-batch reactors in particular. It also discussed the advantages and disadvantages of using batch or semi-batch reactors to produce commercial product. This chapter also distinguished between "process" and "reaction."

REFERENCES

1. Rase H. *Chemical reactor design for process plants: volume one, principles and techniques*. New York, NY: John Wiley & Sons; 1977 [Chapter 12].
2. Schmidt L. *The engineering of chemical reactions*. New York, NY: Oxford University Press; 1998. p. 100–102.
3. Coker A. *Modeling of chemical kinetics and reactor design*. Boston, MA: Gulf Professional Publishing; 2001. p. 221–222.

CHAPTER 2

Batch Reactors

QUANTITATIVE DESCRIPTION OF BATCH REACTORS

Consider the typical batch reactor, which in the laboratory is a spherical glass container with ports for reagent addition and sample removal and possibly a centerline port accommodating an agitator shaft driven by an elevated, i.e., overhead, electric motor. An alternate method of mixing is to place a magnetic stirring bar in the reactor and drive it via a rotating magnet placed beneath the reactor.

We generally do not conduct gas-phase reactions in batch reactors. For a constant liquid volume, constant density reaction or process, the component balance for a laboratory batch reactor is

$$\frac{\partial[\mathrm{A}]}{\partial t}+\left(v_r\frac{\partial[\mathrm{A}]}{\partial r}+\frac{v_\theta}{r}\frac{\partial[\mathrm{A}]}{\partial\theta}+\frac{v_\varphi}{r\sin\theta}\frac{\partial[\mathrm{A}]}{\partial\varphi}\right)$$
$$=D_{\mathrm{AB}}\left(\frac{1}{r^2}\frac{\partial}{\partial r}\left(r^2\frac{\partial[\mathrm{A}]}{\partial r}\right)+\frac{1}{r^2\sin\theta}\frac{\partial}{\partial\theta}\left(\frac{\partial[\mathrm{A}]}{\partial\theta}\right)+\frac{1}{r^2\sin^2\theta}\frac{\partial^2[\mathrm{A}]}{\partial\varphi^2}\right)+R_{\mathrm{A}}$$

where [A] is the concentration of component A (mol/m^3); t is the time (seconds, s); r is the radial direction, generally taken from the geometric center point of the spherical reactor (m); θ is the polar angle taken from the vertical reference axis to the point in question (dimensionless); φ is the azimuthal angle or longitude taken around the reference axis to the point in question (dimensionless); v_r is the velocity in the radial direction (m/s); v_θ is the polar angular velocity (m/s); v_φ is the azimuthal velocity (m/s); D_{AB} is the diffusivity (m^2/s) of component A in solvent B, if a solvent B is present; and, R_{A} is the formation of A or the consumption of A by chemical reaction (mol/m^{3}*s).

We assume the concentration of component A in the r, θ, and φ directions is constant due to agitation and the absence of directed convection[1]; therefore

Batch and Semi-batch Reactors.

$$\left(v_r \frac{\partial[\mathrm{A}]}{\partial r} + \frac{v_\theta}{r} \frac{\partial[\mathrm{A}]}{\partial \theta} + \frac{v_\varphi}{r \sin \theta} \frac{\partial[\mathrm{A}]}{\partial \varphi} \right) = 0$$

If the agitator maintains continuous, thorough mixing of the reactor's contents, then we assume molecular diffusion to be negligible during the chemical reaction; thus

$$D_{\mathrm{AB}} \left(\frac{1}{r^2} \frac{\partial}{\partial r} \left(r^2 \frac{\partial[\mathrm{A}]}{\partial r} \right) + \frac{1}{r^2 \sin \theta} \frac{\partial}{\partial \theta} \left(\frac{\partial[\mathrm{A}]}{\partial \theta} \right) + \frac{1}{r^2 \sin^2\theta} \frac{\partial^2[\mathrm{A}]}{\partial \varphi^2} \right) = 0$$

The component balance for A therefore reduces to

$$\left(\frac{\mathrm{d}[\mathrm{A}]}{\mathrm{d}t} \right)_{\mathrm{Spherical}} = R_{\mathrm{A}}$$

At a production facility, the typical batch reactor is a circular cylinder mounted vertically in a support structure. We generally weld hemispheric caps to the top and bottom of the cylinder. A vertical agitator shaft extends through the top cap of the cylinder. A bearing seal surrounds the vertical shaft as it passes through the top cap, thereby isolating the reactor's contents from the environment. An electric motor powers the agitator shaft via a gearbox. Various nozzles with block valves penetrate the top cap; liquid feeds enter the reactor through these nozzles. The finished product exits the reactor through a valve generally placed at the center of the bottom cap. The component balance for such a reactor is

$$\frac{\partial[\mathrm{A}]}{\partial t} + \left(v_r \frac{\partial[\mathrm{A}]}{\partial r} + \frac{v_\theta}{r} \frac{\partial[\mathrm{A}]}{\partial \theta} + v_z \frac{\partial[\mathrm{A}]}{\partial z} \right)$$
$$= D_{\mathrm{AB}} \left(\frac{1}{r} \frac{\partial}{\partial r} \left(r \frac{\partial[\mathrm{A}]}{\partial r} \right) + \frac{1}{r^2} \frac{\partial^2[\mathrm{A}]}{\partial \theta^2} + \frac{\partial^2[\mathrm{A}]}{\partial z^2} \right) + R_{\mathrm{A}}$$

where [A] is the concentration of component A ($\mathrm{mol/m^3}$); t is the time (s); r is the radial direction, generally taken from the geometric center line of the cylindrical reactor (m); z is the vertical axis of the cylinder (m); θ is the azimuthal angle or longitude taken around the z axis (dimensionless); v_r is the velocity in the radial direction (m/s); v_z is the velocity in the axial direction (m); v_θ is the azimuthal velocity (m/s); D_{AB} is the diffusivity ($\mathrm{m^2/s}$) of component A in solvent B; and, R_{A} is the formation of A or the consumption of A by chemical reaction ($\mathrm{mol/m^3{*}s}$).

As for the laboratory batch reactor, we assume the concentration of component A in the r, θ, and z directions is constant due to agitation and the absence of directed convection.[1] Therefore

$$\left(v_r \frac{\partial[\mathrm{A}]}{\partial r} + \frac{v_\theta}{r}\frac{\partial[\mathrm{A}]}{\partial r} + v_z \frac{\partial[\mathrm{A}]}{\partial z}\right) = 0$$

If the agitator maintains continuous, thorough mixing of the reactor's contents, then we assume molecular diffusion to be negligible during the chemical reaction; thus

$$\left(\frac{1}{r}\frac{\partial}{\partial r}\left(r\frac{\partial[\mathrm{A}]}{\partial r}\right) + \frac{1}{r^2}\frac{\partial^2[\mathrm{A}]}{\partial \theta^2} + \frac{\partial^2[\mathrm{A}]}{\partial z^2}\right) = 0$$

The component balance for a commercial-sized circular, cylindrical batch reactor reduces to

$$\left(\frac{\mathrm{d}[\mathrm{A}]}{\mathrm{d}t}\right)_{\mathrm{Cylinder}} = R_{\mathrm{A}}$$

We can make two statements with regard to

$$\left(\frac{\mathrm{d}[\mathrm{A}]}{\mathrm{d}t}\right)_{\mathrm{Cylinder}} = R_{\mathrm{A}} \quad \text{and} \quad \left(\frac{\mathrm{d}[\mathrm{A}]}{\mathrm{d}t}\right)_{\mathrm{Spherical}} = R_{\mathrm{A}}$$

First, neither rate depends upon a geometric variable, so long as the concentrations and reaction temperature in the cylindrical reactor equal the concentrations in the spherical reactor, which is generally the case for liquid-phase reactions. Thus, upscaling and downscaling a chemical reaction within a batch reactor is relatively straightforward, at least for constant volume, constant density, liquid-phase reactions.[2] If the reaction volume changes or the density of the contained liquid changes during reaction, then we must consider geometric similarity during upscaling and downscaling. Second, both rates depend upon R_{A}.

WHAT IS R_{A}?

To upscale or downscale a batch reactor, we need to know upon what R_{A} depends. Consider the generalized chemical reaction

$$r * \text{Reactants} \rightarrow p * \text{Products}$$

where r and p are the stoichiometric quantities for Reactants and Products, respectively. We know from thermodynamics that, when a chemical reaction reaches equilibrium, the rate of product formation from reactants equals the rate of reactant formation from products. Mathematically

$$K = \frac{k_{\text{Forward}}}{k_{\text{Reverse}}}$$

where K is the equilibrium constant for the chemical reaction; k_{Forward} is a constant describing the rate of product formation from reactants; and, k_{Reverse} is a constant describing the rate of reactant formation from products. We also know from thermodynamics that

$$K = \frac{[\text{Products}]^p}{[\text{Reactants}]^r}$$

$$\frac{k_{\text{Forward}}}{k_{\text{Reverse}}} = \frac{[\text{Products}]^p}{[\text{Reactants}]^r}$$

or

$$k_{\text{Forward}}[\text{Reactants}]^r = k_{\text{Reverse}}[\text{Products}]^p$$

Note that for $k_{\text{Forward}}[\text{Reactants}]^r$, k_{Forward} has units of $(\text{m}^3/\text{mole})^{r-1}(1/\text{s})$. Therefore, if $r = 1$, then

$$(\text{m}^3/\text{mole})^{r-1}(1/s) = (\text{m}^3/\text{mole})^{1-1}(1/s) = (\text{m}^3/\text{mole})^0(1/s) = (1/s)$$

The same is true for $k_{\text{Reverse}}[\text{Products}]^p$. In other words, $k_{\text{Forward}}[\text{Reactants}]^r$ and $k_{\text{Reverse}}[\text{Products}]^p$ have the same units as R_A, which leads to our equating them. Thus, R_A in the forward reaction direction must be a function of reactant concentration while in the reverse reaction direction it must be a function of product concentration. The kinetic theory of gases also indicates that the collision of gas molecules and their resultant reaction with each other depends upon the concentration of each component gas.[3]

Note that r and p in the above equation are the stoichiometric quantities for the reaction after it reaches equilibrium. However, chemical reaction rate studies, i.e., kinetic studies, do not investigate the equilibrium condition of chemical reactions; thermodynamics investigates the equilibrium condition of a chemical reaction.

Kinetics investigates the rate at which a chemical reaction reaches its equilibrium condition. With regard to kinetics, our strongest assertion can only be

$$\frac{d[\text{Reactant}]}{dt} = -k_{\text{Forward}}[\text{Reactant}]^x$$

or

$$\frac{d[\text{Product}]}{dt} = -k_{\text{Reverse}}[\text{Product}]^y$$

where x and y are unknowns not necessarily related to the stoichiometric quantities r and p of the reaction.

The above reaction rate equations indicate that we are eventually going to integrate [Reactant] from $[\text{Reactant}]_{t=0}$ at $t = 0$, when we start the chemical reaction, to $[\text{Reactant}]_t$ at time t, when we stop the chemical reaction. That integration gives us

$$[\text{Reactant}]_t - [\text{Reactant}]_{t=0}$$

which will be negative since

$$[\text{Reactant}]_t << [\text{Reactant}]_{t=0}$$

But, for chemical reactions, rates of change are positive. Therefore, we insert the negative sign in the above equation to obtain a positive rate of change. We interpret the negative sign as denoting consumption of a chemical species; we interpret a positive sign as denoting formation of a chemical species.

Thermodynamics also indicates how the equilibrium constant depends on changing temperature. The Clausius–Clapeyron equation provides that relationship

$$\frac{d(\ln K_{\text{Concentration}})}{dT} = \frac{\Delta U}{RT^2}$$

where $K_{\text{Concentration}}$ is the equilibrium constant based on concentration; ΔU is the internal energy of change required by the chemical reaction; R is the gas constant; and, T is the absolute temperature.[4] Substituting $k_{\text{Forward}}/k_{\text{Reverse}}$ for $K_{\text{Concentration}}$ gives us

$$\frac{\mathrm{d}(\ln(k_{\mathrm{Forward}}/k_{\mathrm{Reverse}}))}{\mathrm{d}T} = \frac{\Delta U}{RT^2}$$

Letting

$$\Delta U = E_{\mathrm{Forward}} - E_{\mathrm{Reverse}}$$

where E_{Forward} is the energy required for the forward reaction to occur and E_{Reverse} is the energy required for the reverse reaction to occur. Upon substituting into the above equation, we obtain

$$\frac{\mathrm{d}(\ln(k_{\mathrm{Forward}}/k_{\mathrm{Reverse}}))}{\mathrm{d}T} = \frac{E_{\mathrm{Forward}} - E_{\mathrm{Reverse}}}{RT^2}$$

Expanding the logarithm term and rearranging the right side of the above equation gives

$$\frac{\mathrm{d}(\ln k_{\mathrm{Forward}})}{\mathrm{d}T} - \frac{\mathrm{d}(\ln k_{\mathrm{Reverse}})}{\mathrm{d}T} = \frac{E_{\mathrm{Forward}}}{RT^2} - \frac{E_{\mathrm{Reverse}}}{RT^2}$$

Equating the terms for the forward reaction and the terms for the reverse reaction yields

$$\frac{\mathrm{d}(\ln k_{\mathrm{Forward}})}{\mathrm{d}T} = \frac{E_{\mathrm{Forward}}}{RT^2}$$

and

$$-\frac{\mathrm{d}(\ln k_{\mathrm{Reverse}})}{\mathrm{d}T} = -\frac{E_{\mathrm{Reverse}}}{RT^2}$$

or

$$\frac{\mathrm{d}(\ln k_{\mathrm{Reverse}})}{\mathrm{d}T} = \frac{E_{\mathrm{Reverse}}}{RT^2}$$

Integrating the equation for the forward reaction gives us

$$\int \mathrm{d}(\ln k_{\mathrm{Forward}}) = \frac{E_{\mathrm{Forward}}}{R}\int \frac{\mathrm{d}T}{T^2}$$

which yields

$$\ln k_{\mathrm{Forward}} = -\frac{E_{\mathrm{Forward}}}{RT} + \ln A$$

where ln A is an integration constant. Expressing the last equation in exponential form yields

$$k_{\text{Forward}} = A\text{e}^{-E_{\text{Forward}}/RT}$$

Note that the units of the pre-exponential factor A must be the same as the units for k_{Forward} since E_{Forward}/RT is dimensionless. A similar equation can be written for the reverse reaction

$$k_{\text{Reverse}} = A\text{e}^{-E_{\text{Reverse}}/RT}$$

Thus, R_{A} not only depends upon the concentration of component A, but it must also depend upon reaction temperature.

ELUCIDATION OF KINETIC RELATIONSHIPS

When investigating the kinetics for a given chemical reaction, we can determine the rate of reaction by measuring the change in reactant concentration or product concentration as functions of time and we can measure each concentration at a particular time: that leaves the rate constant undetermined, unspecified. To completely determine the reaction characteristics of a chemical reaction, we must specify the rate constant also.

Consider the generalized chemical reaction

$$a\text{A} + b\text{B} \rightarrow c\text{C}$$

where a, b, and c are the stoichiometric quantities for molecules A, B, and C, respectively. We know from above that

$$\frac{\text{d[A]}}{\text{d}t} = -k_{\text{Forward}}[\text{A}]^x[\text{B}]^y$$

where [A] is the concentration of reactant A (mol/m^3); [B] is the concentration of reactant B (mol/m^3); t is the time (s); and, k_{Forward} is the rate constant (m^3/mol*s). We refer to x as the reaction order with respect to reactant A and y as the reaction order with respect to reactant B. We define the overall reaction order as

$$n = x + y$$

The concept of reaction order applies only to irreversible reactions or to reactions having a forward reaction rate similar to the equation above. If the reaction is reversible, the reaction order for the forward reaction is generally different from the reaction order for the reverse reaction. Therefore, no overall reaction order exists for reversible reactions since the direction of reaction must be specified when quoting a reaction order.[5]

The above rate equation has three unknowns: k_{Forward}, x, and y. If we assume values for two of the unknowns, then we can mathematically solve this rate equation. However, the validity of our assumption can only be established by experiment. For this example, let us assume, i.e., guess, that $x = 1$ and $y = 0$. This guess implies that the concentration of reactant A changes while the concentration of reactant B does not change since

$$\frac{\mathrm{d[A]}}{\mathrm{d}t} = -k_{\mathrm{Forward}}[\mathrm{A}]^1[\mathrm{B}]^0 = -k_{\mathrm{Forward}}[\mathrm{A}]^1 * 1 = -k_{\mathrm{Forward}}[\mathrm{A}]$$

We can simulate this assumption experimentally by ensuring that $[\mathrm{A}] << [\mathrm{B}]$ at all times during the chemical reaction. In other words, [B] hardly changes as reactant A disappears. This assumption is so common in kinetic investigations that it has a name: the "pseudo-first-order assumption." We write pseudo-first-order rate equations as

$$\frac{\mathrm{d[A]}}{\mathrm{d}t} = -k_{\mathrm{Forward}}^{\mathrm{Pseudo}}[\mathrm{A}]$$

where

$$k_{\mathrm{Forward}}^{\mathrm{Pseudo}} = k_{\mathrm{Forward}}[\mathrm{B}]$$

Rearranging, then integrating the above rate equation gives

$$\int_{\mathrm{A}_{t=0}}^{\mathrm{A}} \frac{\mathrm{d[A]}}{[\mathrm{A}]} = -k_{\mathrm{Forward}}^{\mathrm{Pseudo}} \int_0^t \mathrm{d}t$$

which yields

$$\ln[\mathrm{A}]_t - \ln[\mathrm{A}]_{t=0} = \ln\left\{\frac{[\mathrm{A}]_t}{[\mathrm{A}]_{t=0}}\right\} = -k_{\mathrm{Forward}}^{\mathrm{Pseudo}}t$$

Therefore, if we plot $\ln([\mathrm{A}]_t/[A]_{t=0})$ a function of time, we will obtain a relationship showing $\ln([A]_t/[A]_{t=0})$ decreasing as time increases.

Now consider the decomposition of nitrogen pentoxide, which occurs via the reaction

$$N_2O_5 + N_2O_5 \rightarrow 4NO_2 + O_2$$
$$4NO_2 \Leftrightarrow 2N_2O_4$$

where $\Leftrightarrow$ indicates an equilibrium condition. If the forward reaction of the second equation is slow while the reverse reaction is fast, then we can ignore the equilibrium reaction in our analysis of N_2O_5 decomposition. The rate equation for N_2O_5 decomposition is

$$\frac{d[N_2O_5]}{dt} = -k_{\text{Forward}}[N_2O_5]^x[N_2O_5]^y$$

Note that k_{Forward} has units of m^3/mol*s in the above reaction rate equation. Since N_2O_5 is the starting reagent, we can always conduct the experiment in such a fashion that it is the predominant molecular species in the batch reactor. In other words, only a small fraction of N_2O_5 will decompose. In that case, let us assume $x = 1$ and $y = 0$. The above rate equation then becomes

$$\frac{d[N_2O_5]}{dt} = -k_{\text{Forward}}^{\text{Pseudo}}[N_2O_5]$$

Note that in the above reaction rate equation, $k_{\text{Forward}}^{\text{Pseudo}}$ has units of per seconds. Rearranging and integrating from $[N_2O_5]_{t=0}$ at $t = 0$ to $[N_2O_5]_t$ at time t give us

$$\ln[N_2O_5]_t - \ln[N_2O_5]_{t=0} = \ln\left\{\frac{[N_2O_5]_t}{[N_2O_5]_{t=0}}\right\} = -k_{\text{Forward}}^{\text{Pseudo}}t$$

Table 2.1 contains representative data for the decomposition of N_2O_5 at various times during the chemical reaction.[6] Table 2.1 also contains the ratio $\ln\{[N_2O_5]_t/[N_2O_5]_{t=0}\}$ at various times during the chemical reaction. Figure 2.1 presents a plot of $[N_2O_5]$ versus time. Figure 2.2 presents a plot of $\ln\{[N_2O_5]_t/[N_2O_5]_{t=0}\}$ as a function of time. Visually, the plot suggests a linear relationship between $\ln\{[N_2O_5]_t/[N_2O_5]_{t=0}\}$ and time, as suggested by the last equation above.

With today's computing power and user-friendly software, we can easily determine the relationship between $\ln\{[N_2O_5]_t/[N_2O_5]_{t=0}\}$

Table 2.1 The Decomposition of Nitrogen Pentoxide with Time at 45°C

Time (s)	$[N_2O_5]$ (mmol/L)	$\ln\{[N_2O_5]_t/[N_2O_5]_{t=0}\}$
0	341	0
600	247	− 0.32
1200	185	− 0.61
1800	140	− 0.89
2400	105	− 1.18
3000	78	− 1.48
3600	58	− 1.77
4200	44	− 2.05
4800	33	− 2.34
5400	24	− 2.65
6000	18	− 2.94
7200	10	− 3.53
8400	5	− 4.22
9600	3	− 4.73
∞	0	

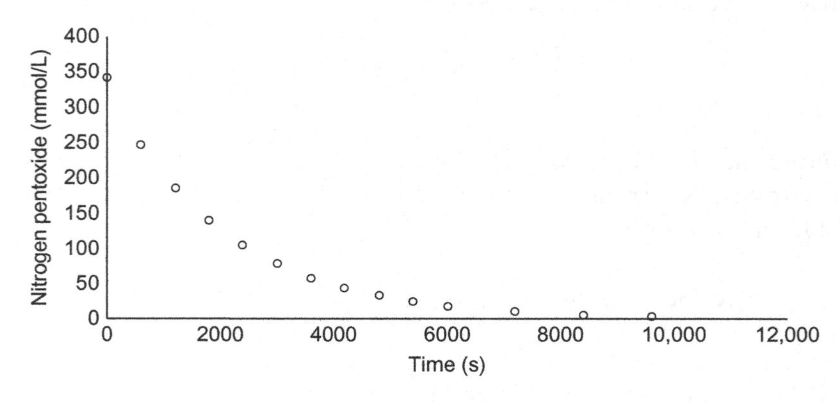

Figure 2.1 Decomposition of N_2O_5 as a function of time at 45° C.

and time. We can even calculate how closely $\ln\{[N_2O_5]_t/[N_2O_5]_{t=0}\}$ approaches that relationship. Figure 2.3 displays a fitted, linear relationship between $\ln\{[N_2O_5]_t/[N_2O_5]_{t=0}\}$ and time; the solid line represents the equation

$$\ln\left\{\frac{[N_2O_5]_t}{[N_2O_5]_{t=0}}\right\} = -(0.0005)t$$

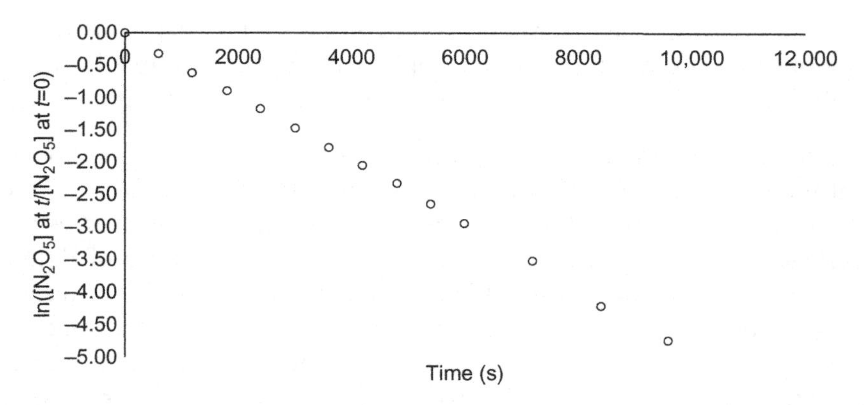

Figure 2.2 Decomposition of N_2O_5 as a function of time at 45° C.

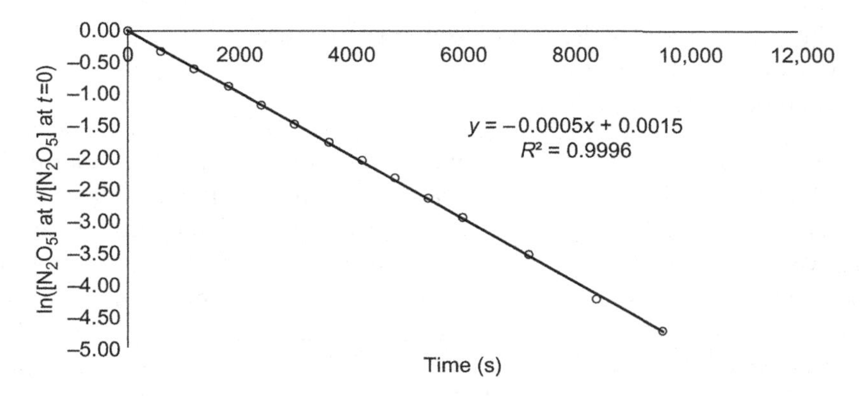

Figure 2.3 Decomposition of N_2O_5 as a function of time at 45° C.

The R^2 for this fitted line is 0.9996. R^2 is the coefficient of determination, which represents

$$\frac{\text{Explained variation}}{\text{Total variation}}$$

Thus, 99.96% of the total variation of the data shown in Figure 2.3 is explained via a linear relationship. The coefficient of regression, or more commonly, the regression coefficient is

$$R = \pm\sqrt{R^2} = \pm\sqrt{\frac{\text{Explained variation}}{\text{Total variation}}}$$

The regression coefficient, R, has the same sign as the slope of the relationship. These definitions are applicable for two variable linear or nonlinear correlations as well as for multiple variable correlation.[7]

But, does Figure 2.3 actually prove our assumption to be correct? Unfortunately, an R^2 of 0.9996 between $\ln\{[N_2O_5]_t/[N_2O_5]_{t=0}\}$ and time does not necessarily indicate causation. A high positive regression coefficient, i.e., R, indicates that the high values of a series of datum entries are associated with the high values of a second series of datum entries. Conversely, a high negative R indicates that the high values of a series of datum entries are associated with the low values of a second series of datum entries.[7] In other words, correlation may occur

- when one variable elicits a response, although it is not the sole causation of the response;
- when two variables depend on a common cause;
- when two variables are interdependent;
- when dependency is fortuitous.

Let us consider the situation when one variable elicits a response, but other variables may actually cause the same response. Figure 2.4 presents a medical study where twenty-two hospital patients received intravenous injections of glucose. Their blood was sampled after a specified time period, then analyzed for glucose. Figure 2.4 shows a linear relationship between injected glucose and retained glucose. R^2 for this correlation is 0.9332. In other words, this correlation explains 93% of the variation in retained glucose. Seven percent of the variation in retained glucose is unexplained by this correlation. There are also a number of other causes that might explain the retained glucose, such

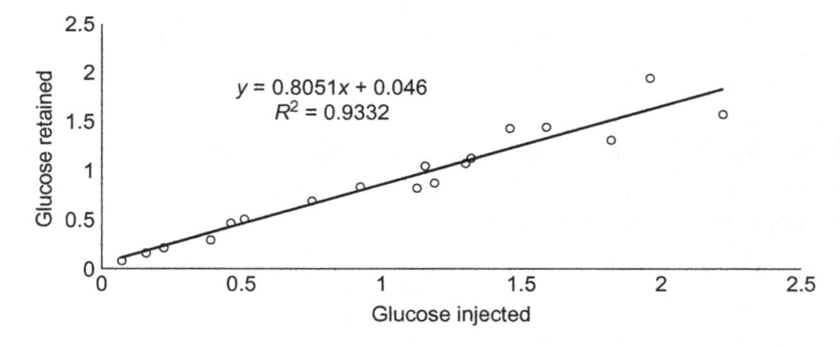

*Figure 2.4 Glucose retained as a function of glucose injected per patient per gram/(kg body weight*hour).*

as some patients were underweight, some were overweight, some were dieting, some were following an intensive exercise regimen, and some were diabetic. Thus, the intravenous glucose injection elicited a response, but other causes could also influence the correlation.[8]

Next, consider Figure 2.5, which shows a linear correlation between the national employment index and the national wholesale price index for the United States from 1919 through 1935. R^2 is 0.6474, which indicates that this correlation explains 65% of the relationship between these two variables. This relationship suggests that when wholesale prices are high, employment is high; and, conversely, when wholesale prices are low, employment is low. This relationship seems reasonable until we realize that both variables depend upon the overall health of the national economy. In other words, when the national economy is "healthy," employment is high and wholesale prices are high; conversely, when the national economy is "unhealthy," employment is low, as are wholesale prices. Thus, Figure 2.5 is an example of two variables depending upon a common cause.[9]

Figure 2.6 shows the relationship between wheat supply in millions of bushels and wheat price in cents/bushel from 1919 through 1933. The correlation is negative; in other words, as wheat production increases, the price of wheat decreases. The correlation is linear and R^2 is 0.5049, indicating that the correlation explains 50% of the variation in the relationship.[10] Figure 2.6 exemplifies an interdependent relationship, i.e., a reciprocal cause and effect relationship. In such relationships, each series, to some extent, is the cause of the other series.[7(p. 127)]

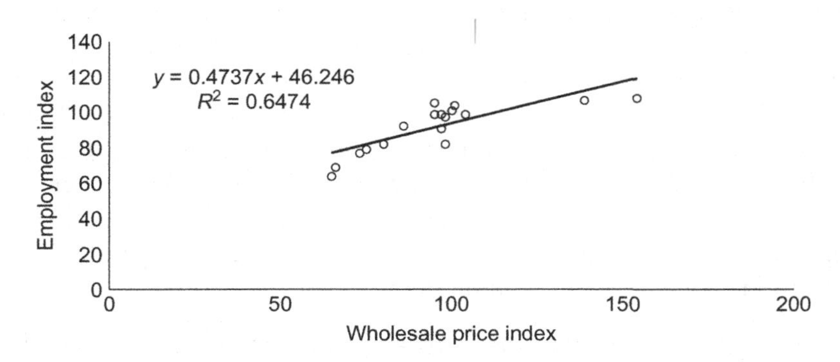

Figure 2.5 Employment index as a function of wholesale price index for 1919 through 1935.

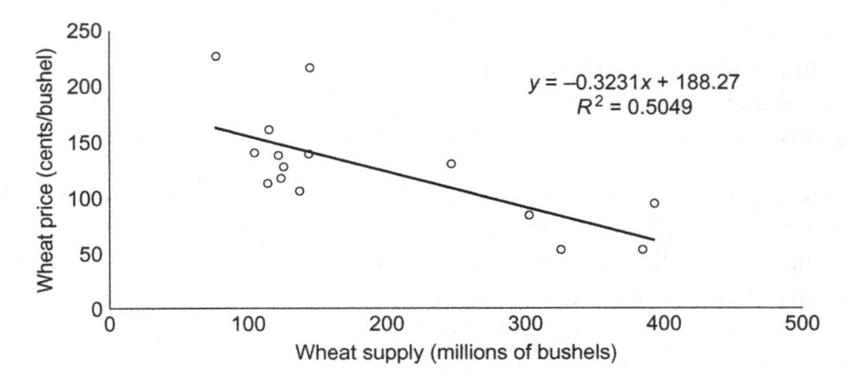

Figure 2.6 Wheat price as a function of wheat supply from 1919 through 1933.

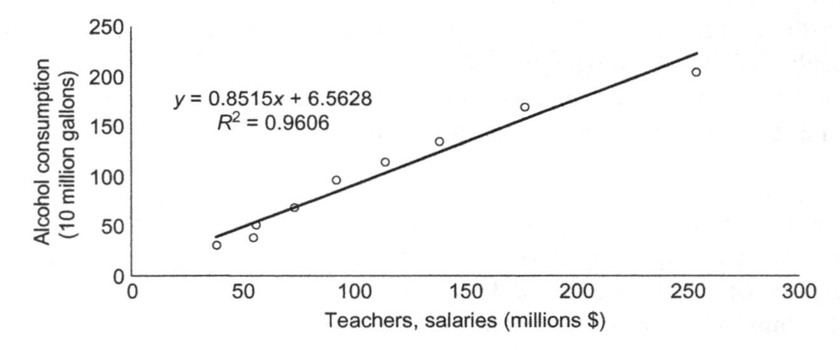

Figure 2.7 Alcohol consumption in the United States as a function of superintendent and teachers' salaries from 1870 through 1910.

Consider Figure 2.7, which shows the consumption of alcohol in the United States, in 10 million gallon increments, as a function of superintendent and teachers' salaries, in million dollar increments, also for the United States. The data is from 1870 through 1910.[11] A correlation of this information suggests the relationship of these two variables is linear with an R^2 of 0.9606. This correlation explains 96% of the variation contained in these two sets of datum. However, does anyone believe that the alcohol consumed in the United States depends so strongly on superintendent and teachers' salaries? Figure 2.7 demonstrates that the correlation of two sets of datum can be fortuitous, i.e., can be independent of cause and effect.

Our original question was: can a linear correlation of two data sets constitute proof of cause and effect? In other words, can we use a linear correlation as confirmation of an assumption in a kinetic investigation? The conservative answer is "no"; the risqué answer is "maybe."

So, why do we continue to use this method when analyzing chemical reaction rate data? We use this method for two reasons. First, it produces an easily integrable differential equation. Second, it only requires one experiment to "prove" or "disprove" the assumption. In reality, that experiment proves or disproves little to nothing. However, we upscale most chemical processes by guessing the reaction order of the rate equation, then "proving" our guess graphically. At best, this procedure of guessing the reaction order of a rate equation is haphazard; at worst, it is dangerous.

Let us reconsider the generalized chemical reaction

$$a\mathrm{A} + b\mathrm{B} \rightarrow c\mathrm{C}$$

The reaction rate equation for this generalized chemical reaction is

$$-\frac{\mathrm{d[A]}}{\mathrm{d}t} = k_{\mathrm{Forward}}[\mathrm{A}]^x[\mathrm{B}]^y$$

Taking the logarithm of this rate equation gives us

$$-\ln\{\mathrm{d[A]}/\mathrm{d}t\} = \ln(k_{\mathrm{Forward}}) + x\ln[\mathrm{A}] + y\ln[B]$$

Rearranging yields

$$-\ln\{\mathrm{d[A]}/\mathrm{d}t\} = \{y\ln[B] + \ln(k_{\mathrm{Forward}})\} + x\ln[\mathrm{A}]$$

The above equation represents a linear relationship between $\ln\{\mathrm{d[A]}/\mathrm{d}t\}$ and $\ln[\mathrm{A}]$ with slope x and intercept $y\ln[\mathrm{B}] + \ln(k_{\mathrm{Forward}})$ for a set of experiments determining $\mathrm{d[A]}/\mathrm{d}t$ as a function of [A]. Note that [B] must be held constant during this set of experiments. We must also perform a set of experiments that measures $\mathrm{d[B]}/\mathrm{d}t$ as a function of [B] while holding [A] constant for the set of experiments. Writing the reaction rate equation in terms of reactant B gives

$$-\frac{\mathrm{d[B]}}{\mathrm{d}t} = k_{\mathrm{Forward}}[\mathrm{A}]^x[\mathrm{B}]^y$$

which, upon taking the logarithm, yields

$$-\ln\{\mathrm{d}[B]/\mathrm{d}t\} = \{x \ln[A] + \ln(k_{\text{Forward}})\} + y \ln[B]$$

In this case, the above equation represents a linear relationship between $\ln\{\mathrm{d}[B]/\mathrm{d}t\}$ and $\ln[B]$ with slope y and intercept $x \ln[A] + \ln(k_{\text{Forward}})$. From these two equations, we can determine x and y and calculate k_{Forward}. Using this procedure, we make no assumptions about the reaction orders x and y. Thus, we are not forced to "prove" our assumption; i.e., guess, concerning x and y graphically.

This procedure requires at least three experiments for each reactant involved in the chemical reaction. Figure 2.8 shows three such curves for the saponification of ethyl acetate at 22°C and 0.1 mol/L NaOH.[12] The reaction is

$$CH_3COOCH_2CH_3 + NaOH \rightarrow CH_3COONa + CH_3CH_2OH$$

Figure 2.8 displays the decrease of ethyl acetate concentration as a function of time. All these experiments were done at 0.1 mol/L NaOH. We must now decide how to calculate $\mathrm{d}[EA]/\mathrm{d}t$ using the information shown in Figure 2.8, where [EA] is ethyl acetate concentration. Historically, we used a straight edge to draw a line tangent to the curve generated by the experiment. We then calculated the slope of that tangent line to obtain, in this case, $\mathrm{d}[EA]/\mathrm{d}t$. However, we also require reactant concentration at the time chosen to determine $\mathrm{d}[EA]/\mathrm{d}t$. The most accurately known reactant concentration is initial reactant concentration. Therefore, we determined the initial $\mathrm{d}[EA]/\mathrm{d}t$ and plotted $\ln\{\mathrm{d}[EA]/\mathrm{d}t|_{t=0}\}$ versus $\ln[A]_{t=0}$. Determining a tangent to

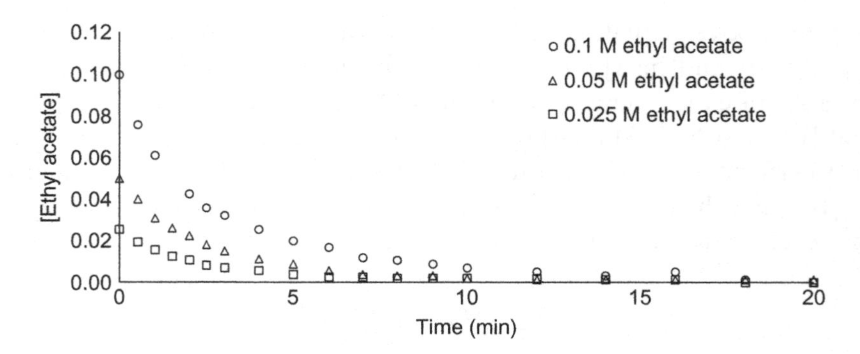

Figure 2.8 [Ethyl Acetate] as a function of time at 22°C. Initial [NaOH] is 0.1 M.

a curve by sight involves significant error, which propagates through the calculations performed during a kinetic analysis. Fortunately, current computing power and user-friendly software obviates such error. Today, we can easily determine the equation describing the relationship between a dependent variable and an independent variable.

Figure 2.9 displays the equations describing the relationships between ethyl acetate concentration and time for the experiments shown in Figure 2.8. The top equation in Figure 2.9 is for an initial 0.1 mol/L ethyl acetate; the middle equation is for an initial 0.05 mol/L ethyl acetate; and, the bottom equation is for an initial 0.025 mol/L ethyl acetate. From these equations, we can calculate the initial reaction rate for ethyl acetate saponification. For 0.1 mol/L ethyl acetate, the equation is

$$[\mathrm{EA}] = 0.0981 - (0.0459)t + (0.0119)t^2 - (0.0017)t^3 + (0.0001)t^4 - (2\times10^{-6})t^5 + (8\times10^{-8})t^6$$

where [EA] is ethyl acetate concentration. Differentiating [EA] with respect to time yields

$$\frac{\mathrm{d}[\mathrm{EA}]}{\mathrm{d}t} = -0.0459 + 2*(0.0119)t - 3*(0.0017)t^2 + 4*(0.0001)t^3 - 5*(2\times10^{-6})t^4 + 6*(8\times10^{-8})t^5$$

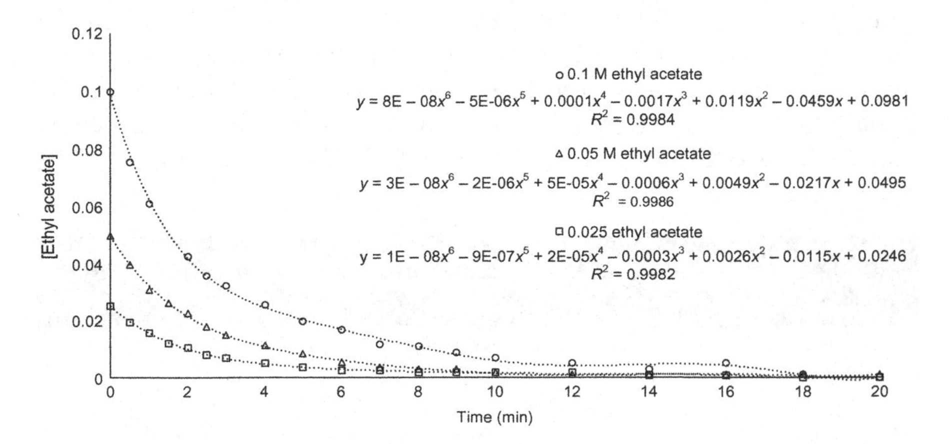

Figure 2.9 [Ethyl Acetate] as a function of time at 22°C. Initial [NaOH] is 0.1 M.

Our interest is in the initial rate of ethyl acetate saponification, which occurs at $t = 0$.[13,14] Therefore, setting $t = 0$ in the above equation gives us

$$\left.\frac{d[EA]}{dt}\right|_{EA=0.1,t=0} = -0.0459$$

Thus, the initial saponification rate for 0.1 mol/L ethyl acetate is 0.046 mol/L*min. Note that we are using the absolute value of $d[EA]/dt$ since chemical reaction rates are positive. The initial saponification rates for 0.05 mol/L and 0.025 mol/L ethyl acetate are easily obtained from the equations shown in Figure 2.9. Those rates are

$$\left.\frac{d[EA]}{dt}\right|_{[EA]=0.05,t=0} = -0.0217 \quad \text{and} \quad \left.\frac{d[EA]}{dt}\right|_{[EA]=0.025,t=0} = -0.0115$$

Table 2.2 tabulates these initial rates and their logarithms, as well as the initial ethyl acetate concentrations and their logarithms. Again, these reaction rates are for an initial 0.1 mol/L NaOH. Figure 2.10 shows the relationship between $\ln\{d[EA]/dt|_{t=0}\}$ and $\ln[EA]_{t=0}$. The linear correlation of the datum points in Figure 2.10 shows the relationship of $\ln\{d[EA]/dt|_{t=0}\}$ and $\ln[EA]_{t=0}$ to be linear with an R^2 of 0.9977. The slope of the correlation is 1.

Figure 2.11 shows the relationship between NaOH concentration and time at various NaOH initial concentrations. Ethyl acetate is initially at 0.1 mol/L for each experiment. Figure 2.11 also displays the equations describing the relationships between NaOH concentration and time at various initial NaOH concentrations. As above, we obtain the initial reaction rates from these equations. Table 2.3 tabulates the initial reaction rates from Figure 2.11 and their logarithms, as well as the initial NaOH concentrations and their logarithms. Figure 2.12

Table 2.2 Initial Reactant Concentration and Initial Reaction Rate for Ethyl Acetate Saponification at Constant NaOH Concentration and Variable Ethyl Acetate Concentration

$[EA]_{t=0}$ (mol/L)	$[NaOH]_{t=0}$ (mol/L)	$d[EA]/dt\|_{t=0}$	$Ln[EA]_{t=0}$	$\ln\{d[EA]/dt\|_{t=0}\}$
0.1	0.1	0.0459	−2.30	−3.08
0.05	0.1	0.0217	−2.99	−3.83
0.025	0.1	0.0115	−3.68	−4.46

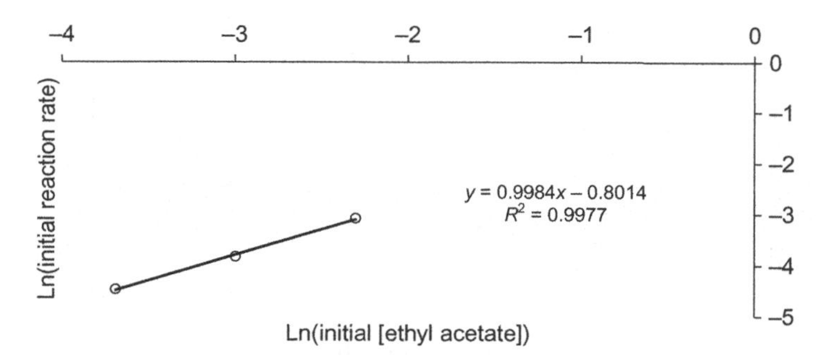

Figure 2.10 Ln(Initial Reaction Rate) as a function of Ln(Initial [Ethyl Acetate]). [NaOH] = 0.1 mol/L, 22° C.

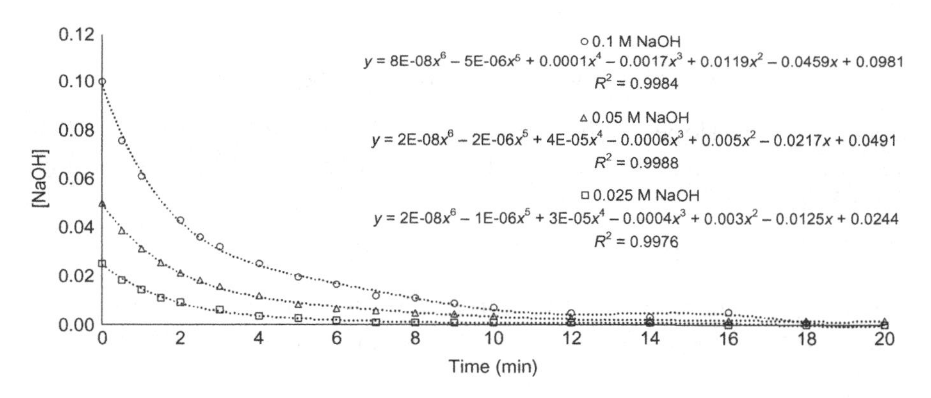

Figure 2.11 [NaOH] as a function of time at 22° C. Initial [EA] is 0.1 M.

Table 2.3 Initial Reactant Concentration and Initial Reaction Rate for Ethyl Acetate Saponification at Constant Ethyl Acetate Concentration and Variable NaOH Concentration

$[EA]_{t=0}$ (mol/L)	$[NaOH]_{t=0}$ (mol/L)	$d[EA]/dt\|_{t=0}$	$Ln[EA]_{t=0}$	$ln\{d[EA]/dt\|_{t=0}\}$
0.1	0.1	0.0459	−2.30	−3.08
0.1	0.05	0.0217	−2.99	−3.83
0.1	0.025	0.0125	−3.68	−4.38

shows the relationship between $\ln\{d[NaOH]/dt|_{t=0}\}$ and $\ln[NaOH]_{t=0}$. The linear correlation of the datum points in Figure 2.12 shows the relationship of $\ln\{d[NaOH]/dt|_{t=0}\}$ and $\ln[NaOH]_{t=0}$ to be linear with an R^2 of 0.9924. The slope of the correlation is 0.94.

With this information, we can calculate k_{Forward} for ethyl acetate saponification at 22°C. The intercept for the reactions conducted at constant initial NaOH concentration is, from Figure 2.10

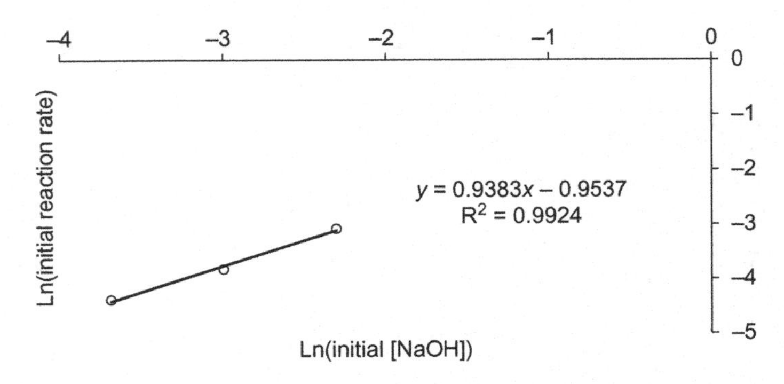

Figure 2.12 Ln(Initial Reaction Rate) as a function of Ln(Initial [NaOH]). [EA] = 0.1 mol/L, 22° C.

$$\text{intercept} = y * \ln[\text{NaOH}]_{t=0} + \ln(k_{\text{Forward}})$$

where y is the slope of the correlation presented in Figure 2.12. The intercept for the reactions conducted at constant initial ethyl acetate is, from Figure 2.12

$$\text{intercept} = x * \ln[\text{EA}]_{t=0} + \ln(k_{\text{Forward}})$$

where x is the slope of the correlation presented in Figure 2.10. Therefore, the equation for the reactions at 0.1 mol/L NaOH, initially, becomes

$$-0.8014 = (0.9383) * \ln(0.1) + \ln(k_{\text{Forward}})$$

which, upon solving yields $k_{\text{Forward}} = 3.9$ L/mol*min. The equation for the reactions at 0.1 mol/L ethyl acetate, initially, becomes

$$-0.9537 = (0.9984) * \ln(0.1) + \ln(k_{\text{Forward}})$$

which, upon solving yields $k_{\text{Forward}} = 3.8$ L/mol*min. We can, therefore, write the reaction rate equation for ethyl acetate saponification as

$$-\frac{\text{d[EA]}}{\text{d}t} = (3.8)[\text{EA}]^1[\text{NaOH}]^{0.94}$$

In this procedure, we determined the reaction order of each reactant by changing initial reactant concentration; thus, the resulting reaction order is with respect to concentration, which is the true reaction order and we give it the symbol n_c, where $n_c = x_c + y_c$.[15–18] Unfortunately, we occasionally find ourselves in a situation where we cannot expend the time or do not have the resources to perform three kinetic experiments per reactant participating in a chemical reaction. In such circumstances, we can determine the reaction order of each reactant and $k_{Forward}$ with regard to time. In this procedure, with regard to ethyl acetate saponification, we perform one experiment at a given ethyl acetate concentration and monitor the decrease of ethyl acetate concentration as a function of elapsed time. Doing so generates one curve similar to the three shown in Figure 2.9. We also perform a similar experiment at a given NaOH concentration and monitor its decrease as a function of time, thereby generating one curve similar to those shown in Figure 2.11. We then calculate the reaction rate for each reactant using the central difference method, which is given by the formula

$$\frac{[A]_{n+2} - [A]_n}{t_{n+2} - t_n} = \text{Rate}$$

where $[A]_n$ is the measured concentration of reactant A at time t_n and $[A]_{n+2}$ is the measured concentration of reactant A at time t_{n+2}, n being a datum index.[12,19]Again, we calculate the rate for early experimental times since we need to specify the concentrations of all co-reactants in order to calculate $k_{Forward}$, as done above. We then correspond each such calculated rate with the reactant concentration at time t_{n+1}; namely, $[A]_{n+1}$, and convert each pair to its logarithm. Finally, we plot ln(Rate) as a function of ln[A]. For ethyl acetate saponification, A is either ethyl acetate or NaOH. Figures 2.13 and 2.14 present ln(Rate) as a function of ln[EA] and ln[NaOH],

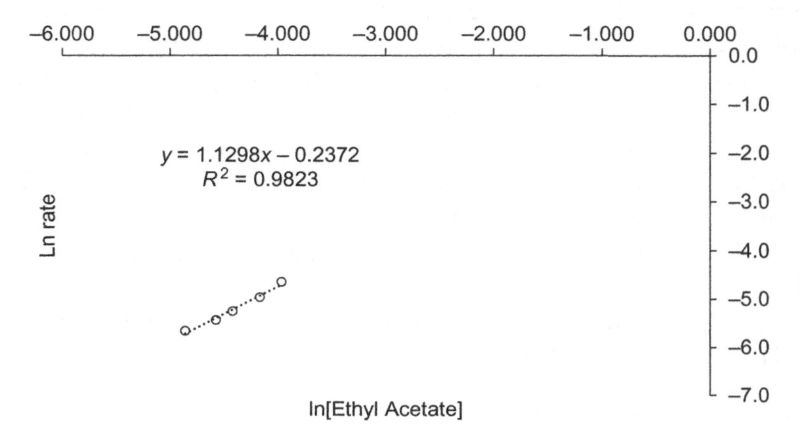

Figure 2.13 Order of reaction, Nt, for ethyl acetate.

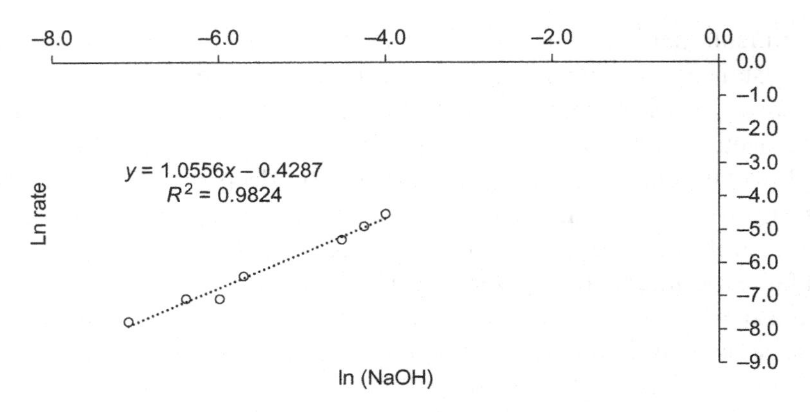

Figure 2.14 Order of reaction, Nt, for NaOH, 22°C.

respectively, where ln(Rate) has been calculated by the central difference method. Figure 2.13 shows that the reaction order with respect to ethyl acetate concentration is 1.1. Figure 2.14 shows that the reaction order with respect to NaOH concentration is also 1.1. We can, therefore, write the reaction rate equation as

$$\frac{d[EA]}{dt} = k_{Forward}[EA]^{1.1}[NaOH]^{1.1}$$

From Figures 2.13 and 2.14, we calculate $k_{Forward}$ to be 9.0 and 7.4 L/mol*min, respectively. The average for $k_{Forward}$ is 8.2 L/mol*min.

We can also calculate the reaction order with respect to time using the polynomials displayed in Figures 2.9 and 2.11. Differentiating a chosen polynomial with respect to time yields the reaction rate for a set of given reactant concentrations. Then, substituting a set of given times into the original polynomial and the reaction rate polynomial and taking the logarithms of the calculated results provides us a set of ln(Rate) and ln[A] data, where [A] is either ethyl acetate or NaOH concentration. Again, we are interested in early reaction times, i.e., in the initial reaction rate period. Using this procedure for $[EA]_{t=0}$ of 0.1 mol/L and $[NaOH]_{t=0}$ of 0.1 mol/L and for $[EA]_{t=0}$ of 0.1 mol/L and $[NaOH]_{t=0}$ of 0.025, both from Figure 2.9, we get reaction orders with respect to time of 1.2 for ethyl acetate concentration and 1.1 for NaOH concentration. The reaction rate equation is then

$$\frac{d[EA]}{dt} = k_{Forward}[EA]^{1.2}[NaOH]^{1.1}$$

Table 2.4 Reaction Order, $k_{Forward}$, and Reaction Rate per Experimental Procedure

Experimental Procedure	Ethyl Acetate Reaction Order	NaOH Reaction Order	$k_{Forward}$ (L/mol*s)	Reaction Rate (mol/L*s)
Initial concentration	1	0.94	3.8	$-\frac{d[EA]}{dt}=(3.8)[EA]^{1}[NaOH]^{0.94}$
Initial time central difference	1.1	1.1	8.2[a]	$-\frac{d[EA]}{dt}=(8.2)[EA]^{1.1}[NaOH]^{1.1}$
Initial time polynomial	1.2	1.1	8.8[a]	$-\frac{d[EA]}{dt}=(8.8)[EA]^{1.2}[NaOH]^{1.1}$

[a] *Average of calculated results.*

In this procedure, $k_{Forward}$ is 7.9 L/mol*min from the [EA] = [NaOH] = 0.1 mol/L and it is 9.8 L/mol*min from the [EA] = 0.1 mol/L and [NaOH] = 0.025 mol/L experiment. The average for $k_{Forward}$ is 8.8 L/mol*min.

In the above two procedures, we determine the reaction order of each reactant with respect to time; therefore, we denote these reaction orders as $n_t = x_t + y_t$.[15–18]

We have now determined the reaction order of each reactant for ethyl acetate saponification via two procedures: one, with respect to initial reactant concentration; the other, with respect to initial reaction time. We have a choice with regard to the latter procedure: we can calculate the reaction rate using a central difference algorithm or we can generate a polynomial equation describing the change in reactant concentration as a function of time, then differentiate that polynomial to obtain the reaction rate. Table 2.4 presents the results for each of these procedures for ethyl acetate saponification. So—which reaction rate equation is correct? The more appropriate question is: which reaction rate equation is valid? The most valid reaction rate equation is the one based on initial reactant concentrations. The reaction rate equations based on initial reaction time are approximations of it. The sets of equations will be similar but not necessarily the same since we determined each set via a different experimental procedure.

SUMMARY

This chapter presented a quantitative description of batch reactors. It also "derived" the variables which control a chemical reaction conducted in a batch reactor. This chapter included a discussion about specifying the proportionality constant, i.e., the reaction rate constant, which quantifies the relationship between chemical reaction rate and concentration. This chapter used the saponification of ethyl acetate to demonstrate the various procedures for specifying the reaction rate constant.

REFERENCES

1. Brotz W. *Fundamentals of chemical reaction engineering*. Reading, MA: Addison-Wesley Publishing Company, Inc.; 1965. p. 216–219.
2. Harriott P. *Chemical reactor design*. New York, NY: Marcel Dekker, Inc.; 2003. p. 82.
3. Kauzmann W. *Kinetic theory of gases*. New York, NY: W. A. Benjamin, Inc.; 1966 [Chapter 5].
4. Rose J. *Dynamic physical chemistry: a textbook of thermodynamics, equilibria and kinetics*. London, UK: Sir Isaac Pitman and Sons, Ltd; 1961. p. 376–389.
5. E. Nauman, *Chemical reactor design, optimization, and scaleup*. 2nd ed. New York, NY: John Wiley & Sons, Inc.; 2008. p. 10.
6. Daniels F. *Chemical kinetics*. Ithaca, NY: Nabu Public Domain Reprints originally published by Cornell University Press; 1938. p. 63.
7. Croxton F. *Elementary statistics with applications in medicine and the biological sciences*. New York, NY: Dover Publications, Inc.; 1959. p. 120–126.
8. Cain J, Belk W. The assimilation rate of intravenously injected glucose in hospital patients. *The American Journal of the Medical Sciences* 1942;**203**(3):359–63.
9. Anon. *Handbook of labor statistics*. Washington, DC: U.S. Government Printing Office; 1936. p. 132 and 673.
10. Anon. *Yearbook of agriculture*. Washington, DC: U.S. Government Printing Office; 1935. p. 363–364.
11. Anon. *Statistical abstract of the United States*. Washington, DC: U.S. Government Printing Office; 1918. p. 830 and 835.
12. Bor-Yea Hsu, University of Houston, personal communication, August 2014.
13. Hall K, Quickenden T, Watts D. Rate Constants from Initial Concentration Data. *Journal of Chemical Education* 1976;**53**:493.
14. Espensen J. *Chemical kinetics and reaction mechanisms*. 2nd ed. New York, NY: McGraw-Hill; 1995. p. 8.
15. Letort M. Definition and Determination of the Two Orders of a Chemical Reaction. *Journal de Chimie Physique* 1937;**34**:206 quoted from:Benson S. *The foundations of chemical kinetics*. New York, NY: McGraw-Hill; 1960. p. 379, 380, 381, 383.
16. Letort M. Contribution a l'etude du mecanisme de la reaction chimique. *Bulletin de la Societe Chimique de France* 1942;**9**:1 quoted from:Benson S. *The foundations of chemical kinetics*. New York, NY: McGraw-Hill; 1960. p. 379, 380, 381, 383.
17. Laidler K. *Chemical kinetics*. 2nd ed. New York, NY: McGraw-Hill; 1965. p. 15–17.
18. Laidler K. *Reaction kinetics: volume 1, homogeneous gas reactions*. Oxford, UK: Pergamon Press; 1963. p. 16–19.
19. Maxwell Smith, University of Houston personal communication, August 2014.

CHAPTER 3

Scaling Batch Reactors

SCALING CONCENTRATION, TEMPERATURE, VOLUME, AGITATION

Our interest in chemical reaction kinetics arises from our need to use large-sized batch reactors to produce enough product to satisfy market demand. There is another reason for using large batch reactors: as reactor size increases, manufacturing cost decreases to a minimum. Management enjoys the sound of that proposition; therefore, they demand that product be produced in as large a batch reactor as possible.

Figure 3.1 shows a typical curve for the formation of product in a laboratory-sized batch reactor. In this case, the product is acetate from the saponification of ethyl acetate[1]

$$CH_3COOC_2H_5 + NaOH \rightarrow CH_3COO^-Na^+ + HOC_2H_5$$

To saponify ethyl acetate in commercially viable quantities, we need to use larger-sized batch reactors, which means reproducing the curve in Figure 3.1 using data from each successively larger batch reactor.

Historically, the CPI from its inception until the mid-1970s did just that when developing a new chemical process: it built successively larger batch reactors, learned to operate them, gathered operating and future design information from them, then used that information to design the next larger batch reactor. Eventually, the process was declared "commercial" and all subsequently built batch reactors were a copy of the one used during the final stage of process development. This process development procedure was the only one available before the advent of inexpensive computation. To develop a process via this procedure requires significant time and considerable capital, both of which are scarce commodities in today's global economy.

Since the mid-1950s, the cost of computation via digital computers has decreased annually. With the invention of the personal computer

Batch and Semi-batch Reactors.

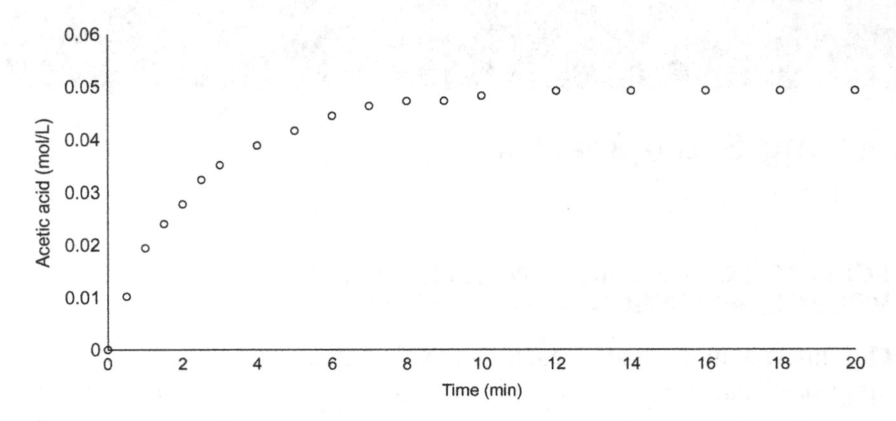

Figure 3.1 Formation of acetic acid by saponification of ethyl acetate, 22°C.

and user-friendly software since the mid-1980s, the cost of computation has become minuscule. The CPI has embraced computation as a means for modifying its historic process development procedure. Today, we use reaction data gathered from a laboratory-sized batch reactor to model as large a batch reactor as deemed safe to commission and operate based on that laboratory data.

To simulate, or model, a batch reactor via computation, we first derive the differential equation

$$\frac{\mathrm{d}[\mathrm{P}]}{\mathrm{d}t} = R_{\mathrm{P}}$$

where [P] is the concentration of product P (mol/m^3) and R_{P} is the formation of P by chemical reaction (mol/m^{3}*s or mol/m^{3}*min). Our task is to determine R_{P} by one of four methods. Those methods are

1. guess the concentration dependence of R_{P}, solve the above differential equation analytically, plot the resulting dependent variable as a function of time, then calculate k, the rate constant, from the slope of the line;
2. determine the concentration dependence of R_{P} using the initial concentration experimental method, then calculate k;
3. determine the concentration dependence of R_{P} using the initial time central difference method, then calculate k;
4. determine the concentration dependence of R_{P} using the initial time polynomial method, then calculate k.

Table 2.4 contains the rate constant for ethyl acetate saponification as determined by the latter three methods above.

Ethyl acetate saponification is a much studied reaction. If we assume that R_P depends linearly on ethyl acetate concentration and linearly on NaOH concentration, i.e., if we follow the first procedure listed above, then we can write the above differential equation as

$$-\frac{d[EA]}{dt} = -\frac{d[NaOH]}{dt} = k[EA][NaOH] = \frac{d[P]}{dt}$$

where all bracketed terms indicate units of mol/L since the volume of the fluid in the reactor is constant, as is the density of the fluid in the reactor. The above differential equation is generally written in terms of extent of reaction x, which gives

$$\frac{d[P]}{dt} = \frac{dx}{dt} = k([EA]_{t=0} - x)([NaOH]_{t=0} - x)$$

The solution to this differential equation is

$$\frac{1}{[EA]_{t=0} - [NaOH]_{t=0}} \ln\left\{\frac{[EA]_{t=0}([NaOH]_{t=0} - x)}{[NaOH]_{t=0}([EA]_{t=0} - x)}\right\} = kt$$

Rearranging the above equation gives

$$\ln\left\{\frac{[EA]_{t=0}([NaOH]_{t=0} - x)}{[NaOH]_{t=0}([EA]_{t=0} - x)}\right\} = ([EA]_{t=0} - [NaOH]_{t=0})kt$$

Therefore, plotting $\ln\{[EA]_{t=0}([NaOH]_{t=0} - x)/[NaOH]_{t=0}([EA]_{t=0} - x)\}$ as a function of time t yields a straight line if our initial guess is correct. The slope of this straight line is

$$([EA]_{t=0} - [NaOH]_{t=0})k$$

From which we calculate k. This method, the first one listed above, yields 8.3 L/mol*min or 0.14 L/mol*s for the ethyl acetate saponification rate constant k.[2] As shown in Chapter 2, the initial time central difference method gives 8.3 L/mol*min and the initial time polynomial method gives 8.8 L/mol*min and the initial concentration method yields 3.8 L/mol*min for the ethyl acetate saponification rate constant (see Table 2.4).

To generate a curve similar to Figure 3.1, we must solve

$$\frac{1}{[\mathrm{EA}]_{t=0} - [\mathrm{NaOH}]_{t=0}} \ln\left\{\frac{[\mathrm{EA}]_{t=0}([\mathrm{NaOH}]_{t=0} - x)}{[\mathrm{NaOH}]_{t=0}([\mathrm{EA}]_{t=0} - x)}\right\} = kt$$

for x, which, in this case, is either acetic acid concentration or ethanol concentration. We will choose acetic acid concentration. Solving the above equation for x gives

$$x = \frac{[\mathrm{NaOH}]_{t=0}(1 - \mathrm{e}^{([\mathrm{NaOH}]_{t=0} - [\mathrm{EA}]_{t=0})kt})}{(1 - [\mathrm{NaOH}]_{t=0}/[\mathrm{EA}]_{t=0})\mathrm{e}^{([\mathrm{NaOH}]_{t=0} - [\mathrm{EA}]_{t=0})kt}}$$

With today's computing power and software packages, we can easily calculate x, so long as we know k, as a function of time for any given $[\mathrm{EA}]_{t=0}$ and $[\mathrm{NaOH}]_{t=0}$. Doing so for $[\mathrm{EA}]_{t=0} = 0.05$ mol/L and $[\mathrm{NaOH}]_{t=0} = 0.10$ mol/L and for each of the ks determined above yields Figure 3.2, which also contains the experimental results for acetate formation.[1]

Figure 3.2 shows that none of the computational simulations match the experimental data well. The simulation using a k of 3.8 L/mol*min predicts a low outcome during the first half of the reaction while the two simulations using a k of 8.3 and 8.8 L/mol*min predict high outcomes during the first half of the reaction. Note that the simulations using 8.3 and 8.8 L/mol*min overlap each other during the reaction

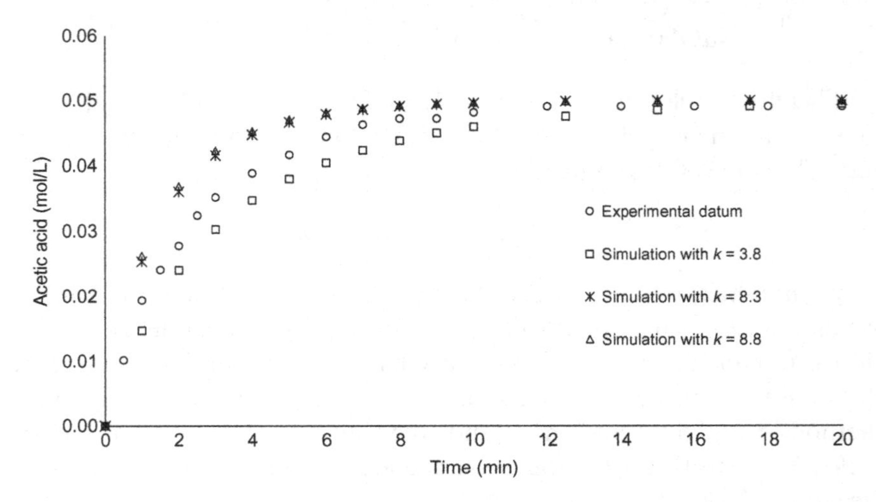

Figure 3.2 Formation of acetic acid by saponification of ethyl acetate. [Acetic Acid] = 0.05 mol/L; [NaOH] = 0.10 mol/L, 22° C.

period. The outcomes shown in Figure 3.2 may be acceptable for low energetic, innocuous, simple reactions, such as ethyl acetate saponification. But, unfortunately, most chemical reactions of interest are complex, energetic, and occasionally spiteful. This method does not work well for such reactions.

However, today, with little effort, we can obtain a best fit equation describing the relationship between a dependent variable and one or more independent variables. For example, consider the best fit polynomial for the experimental data displayed in Figure 3.3. That polynomial is

$$[\mathrm{AA}] = 0.0005 + 0.0217\,t - 0.0049\,t^2 + 0.0006\,t^3 - (5\times10^{-5})t^4 + (2\times10^{-6})t^5 - (3\times10^{-8})t^6$$

where [AA] is acetic acid concentration in mol/L. The above equation is the integrated form of the differential equation describing a chemical reaction in a batch reactor, namely

$$\frac{\mathrm{d}[\mathrm{AA}]}{\mathrm{d}t} = R_{\mathrm{AA}} = 0.0217 - 2(0.0049)t + 3(0.0006)t^2 - 4(5\times10^{-5})t^3 + 5(2\times10^{-6})t^4 - 6(3\times10^{-8})t^5$$

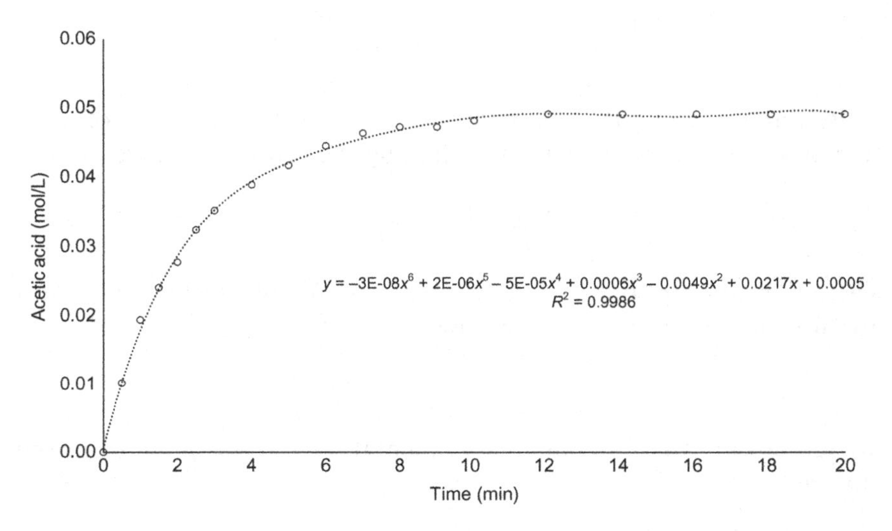

Figure 3.3 Formation of acetic acid by saponification of ethyl acetate. Initially 0.05 mol/L ethyl acetate; 0.1 mol/L NaOH, 22°C.

In this case, we write R_{AA} in terms of time rather than in terms of concentration directly and temperature indirectly through the Arrhenius equation

$$k = A\mathrm{e}^{-(E/RT)}$$

The question is: can we use an equation such as

$$[\mathrm{AA}] = 0.0005 + 0.0217\,t - 0.0049\,t^2 + 0.0006\,t^3 - (5\times10^{-5})t^4 + (2\times10^{-6})t^5 - (3\times10^{-8})t^6$$

to design a batch reactor? The answer is "yes" if we know how the coefficients of the polynomial depend upon concentration, temperature, agitation, and reactive volume.

We know from experience that chemical reaction rate depends upon reactant concentration; thus, in our case

$$\frac{\mathrm{d}[\mathrm{AA}]}{\mathrm{d}t} \propto [\mathrm{EA}]^n[\mathrm{NaOH}]^m$$

where [AA] represents acetic acid molar concentration and [EA] and [NaOH] represent ethyl acetate and NaOH molar concentrations, respectively. We can remove the proportionality sign by assuming a proportionality constant k, which is the reaction rate constant. The above equation then becomes, at $t = 0$

$$\left.\frac{\mathrm{d}[\mathrm{AA}]}{\mathrm{d}t}\right|_{t=0} = k[\mathrm{EA}]^n_{t=0}[\mathrm{NaOH}]^m_{t=0}$$

By differentiating the polynomial describing the formation of acetic acid as a function of time, then evaluating the rate at $t = 0$, we obtain

$$\left.\frac{\mathrm{d}[\mathrm{AA}]}{\mathrm{d}t}\right|_{t=0} = \beta_{[\mathrm{EA}]_{t=0}}$$

where the subscript on [EA] denotes the initial ethyl acetate concentration for that experiment. Combining the last two equations yields

$$\beta_{[\mathrm{EA}]_{1,t=0}} = k[\mathrm{EA}]^n_{1,t=0}[\mathrm{NaOH}]^m_{1,t=0}$$

where the subscript 1 denotes experimental set 1. Taking the logarithm of it gives

$$\ln\left(\beta_{[\mathrm{EA}]_{1,t=0}}\right) = n\ln[\mathrm{EA}]_{1,t=0} + m\ln[\mathrm{NaOH}]_{1,t=0} + \ln k$$

At a different ethyl acetate concentration, denoted by $[EA]_2$, but the same NaOH concentration, i.e., $[NaOH]_{1,t=0}$, we get the equation

$$\ln\left(\beta_{[EA]_{2,t=0}}\right) = n\ \ln[EA]_{2,t=0} + m\ \ln[NaOH]_{1,t=0} + \ln k$$

Subtracting the latter equation from the former equation gives us

$$\ln\left(\beta_{[EA]_{1,t=0}}\right) - \ln\left(\beta_{[EA]_{2,t=0}}\right) = n\ \ln[EA]_{1,t=0} - n\ \ln[EA]_{2,t=0} + m\ \ln[NaOH]_{1,t=0} + \ln k - m\ \ln[NaOH]_{1,t=0} - \ln k$$

which reduces to

$$\ln\left(\beta_{[EA]_{1,t=0}}\right) - \ln\left(\beta_{[EA]_{2,t=0}}\right) = n\ \ln[EA]_{1,t=0} - n\ \ln[EA]_{2,t=0}$$

Rearranging yields

$$\ln\left(\frac{\beta_{[EA]_{1,t=0}}}{\beta_{[EA]_{2,t=0}}}\right) = n\ \ln\left(\frac{[EA]_{1,t=0}}{[EA]_{2,t=0}}\right)$$

Taking the anti-logarithm of this equation gives

$$\frac{\beta_{[EA]_{1,t=0}}}{\beta_{[EA]_{2,t=0}}} = \left(\frac{[EA]_{1,t=0}}{[EA]_{2,t=0}}\right)^n$$

If $n = 1$, as is the case for ethyl acetate saponification, we find

$$\frac{\beta_{[EA]_{1,t=0}}}{\beta_{[EA]_{2,t=0}}} = \frac{[EA]_{1,t=0}}{[EA]_{2,t=0}}$$

Thus, when $n = 1$, the ratio of the polynomial coefficients equals the ratio of the initial reactant concentrations for two given experiments of the same chemical reactions, provided all other variables are held constant.

Figure 3.4 shows ethyl acetate concentration as a function of time for three initial concentrations. Figure 3.4 also displays the best fit polynomial for each reaction. Differentiating each polynomial, then evaluating the resulting derivative at $t = 0$, yields the initial rates for each initial concentration. Those initial concentrations and initial reaction rates are

$[EA]_{t=0}$	$\beta_{t=0}$
0.10	0.0495
0.05	0.0217
0.025	0.0115

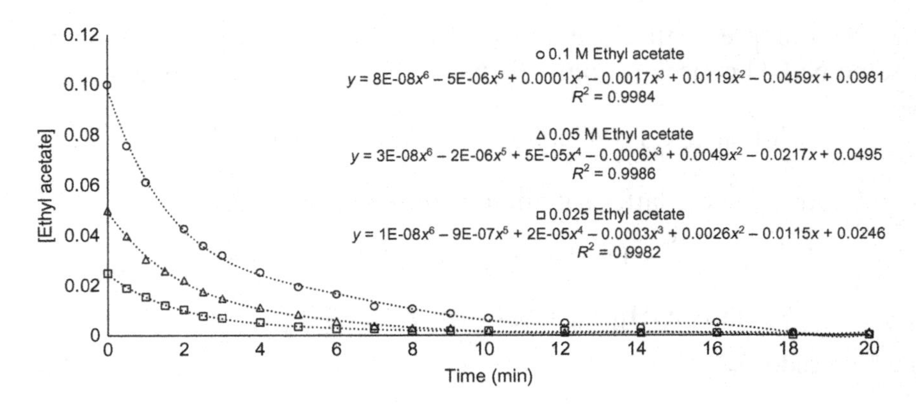

Figure 3.4 [Ethyl Acetate] as a function of time at 22° C. Initial [NaOH] is 0.1 M.

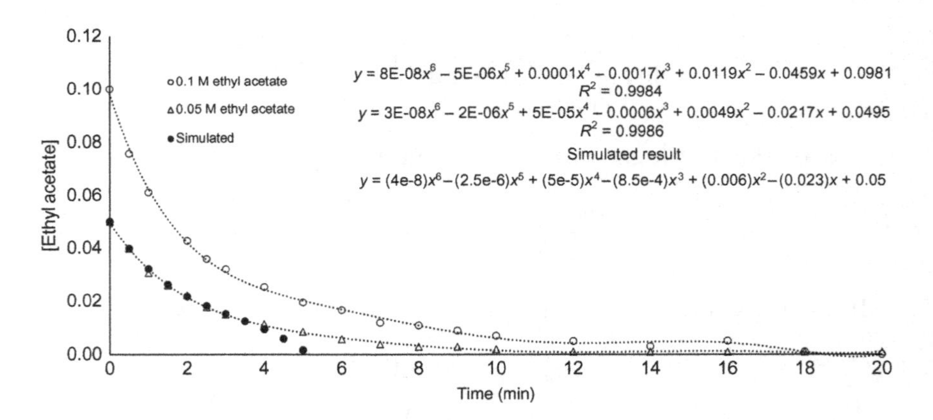

Figure 3.5 [Ethyl Acetate] as a function of time at 22° C. Initial [NaOH] is 0.1 M.

Thus, for a first-order reactant, the initial reaction rate can be estimated using

$$\beta_{[\mathrm{R}]_{1,t=0}} = \left(\frac{[\mathrm{R}]_{1,t=0}}{[\mathrm{R}]_{2,t=0}}\right)\beta_{[\mathrm{R}]_{2,t=0}}$$

where [R] denotes reactant concentration. This result offers us the opportunity to generate similar curves for different, untested reactant concentrations. Consider Figure 3.5: it displays the polynomial $[\mathrm{EA}]_{t=0}$ of 0.10 mol/L, which is

$$[\mathrm{EA}] = (8 \times 10)x^6 - (5 \times 10^{-6})x^5 + (0.0001)x^4 - (0.0017)x^3 + (0.0119)x^2 - 0.0495x + 0.0981$$

where x denotes time. Since the initial reaction rate of ethyl acetate saponification depends linearly on [EA], we can use the above equation to establish a similar equation for an initial 0.05 mol ethyl acetate/L. That equation is

$$[\mathrm{EA}] = (4 \times 10)x^6 - (2.5 \times 10^{-6})x^5 + (5 \times 10^{-5})x^4 - (8.5 \times 10^{-4})x^3 + (0.00595)x^2 - 0.023x + 0.05$$

Figure 3.5 shows the simulated ethyl acetate concentration as a function of time and compares it to the experimentally determined curve for an initial 0.05 mol/L. The two curves overlap for the initial reaction period. However, the simulation indicates complete ethyl acetate consumption after 5 min, whereas the experimental data indicates the presence of a miniscule amount of ethyl acetate in the reactor. We can, therefore, use this method to simulate the most active period of a reaction in a batch reactor.

The next question is: how do the coefficients of a polynomial describing reactant or product concentration as a function of time change with process temperature? Thermodynamics suggests that the reaction rate constant is proportional to the exponential of inverse process temperature T, namely

$$k = A\mathrm{e}^{-E/RT}$$

where A is a proportionality constant and T is expressed as absolute temperature. If we plan to scale a batch process using a regressed polynomial equation, then each of the coefficients of the polynomial must respond to changes in process temperature. Consider the generalized polynomial equation for product formation

$$[\mathrm{P}] = \alpha + \beta t - \gamma t^2 + \delta t^3 - \varepsilon t^4 + \zeta t^5 - \eta t^6$$

where P denotes product. α is the initial product concentration; thus, it will be zero at the start of the reaction, unless product forms before the first sample is taken. β is the initial reaction rate which must relate to process temperature as

$$\beta = A\mathrm{e}^{-E/RT}$$

The other polynomial coefficients must also respond to changes in process temperature. The simplest assumption is they relate exponentially to process temperature. In other words, an equation similar to the one above can be used to relate each polynomial coefficient to process temperature.

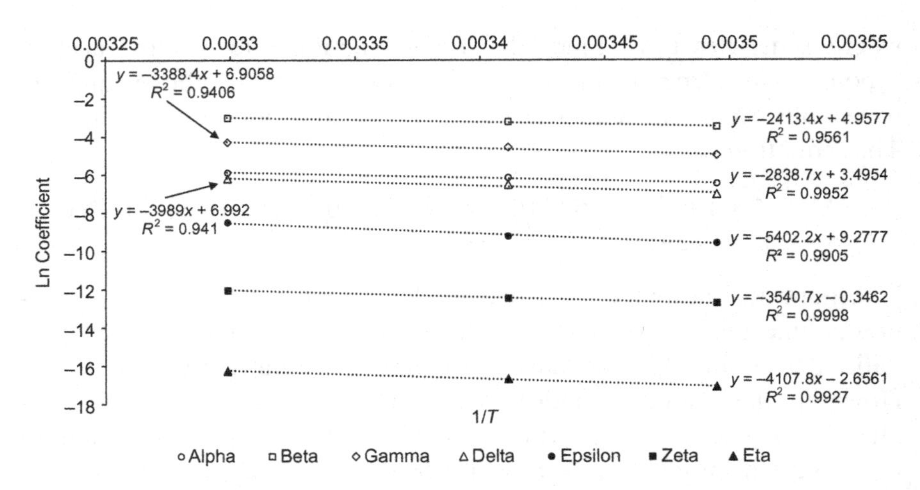

Figure 3.6 Temperature dependence of polynomial coefficients for acetic acid formation via saponification, 0.1 M ethyl acetate + 0.1 M NaOH.

Figure 3.6 plots ln(polynomial coefficient) as a function of $1/T$ for the formation of acetic acid, i.e., acetate, via NaOH saponification. Initially, ethyl acetate was 0.1 mol/L, as was NaOH. Saponification was conducted at 286, 293, and 303 K. Figure 3.5 shows that each polynomial coefficient can be related to process temperature via an equation such as

$$X = A_X e^{-E_X/RT}$$

where X represents α, β, λ, δ, ε, ζ, and η individually. Note that A_X and E_X are most likely dependent on the polynomial coefficient. Note that for product formation, α should equal zero at $t = 0$; however, fitting product formation as a polynomial function of time shows that some reaction occurs upon immediately mixing "at temperature" reactants. Thus, ln(α) demonstrates a slight dependence on $1/T$(K), as displayed in Figure 3.6. Figure 3.6 also shows that the slope of the Arrhenius relation increases as the power on time increases until t^4, after which the slope fluctuates. Thus, the "energy of activation" associated with each power coefficient increases to and including the fourth power on time; thereafter, the energy of activation fluctuates.

Now consider sucrose inversion, which is

$$\text{Sucrose} + \text{HCl} \rightarrow \text{Glucose} + \text{Fructose}$$

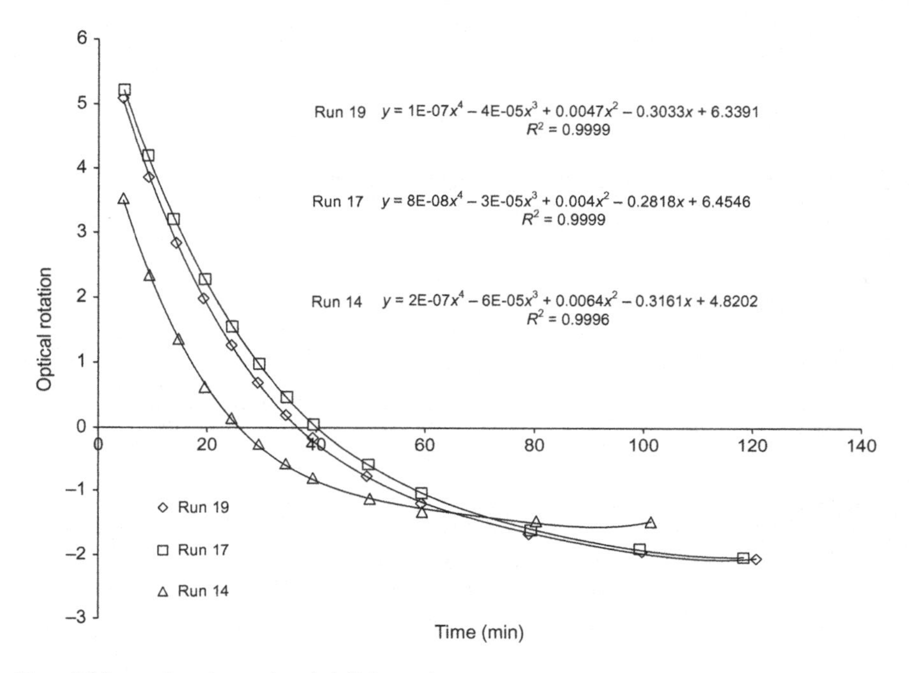

Figure 3.7 Sucrose inversion monitored via light rotation.

We monitor the progress of sucrose inversion with a polarimeter, which measures the optical rotation of light passing through a sample. Sucrose is dextrorotary, i.e., it rotates light rightward. Glucose and fructose mixtures are slightly levorotary, i.e., they rotate light leftward. In this reaction, sucrose disappears and glucose and fructose form, thus light rotation rightward decreases until eventually the light is completely levorotary. Figure 3.7 displays the progress of three typical sucrose inversions monitored by optical rotation.[3]

The information in Figure 3.7 can be converted to molar concentrations, after which we can apply the usual mathematical methods used in a chemical kinetics investigation. Figure 3.8 shows sucrose concentration as a function of time for Run 19. The polynomial equation fitting sucrose concentration to time is displayed in Figure 3.8. Note that it has the general form

$$[\mathrm{R}] = \alpha - \beta t + \gamma t^2 - \delta t^3 + \varepsilon t^4 - \zeta t^5 + \eta t^6$$

where R denotes reactant. α is the initial feed concentration; thus, it will be independent of process temperature. β is the initial reaction rate which must relate to process temperature as

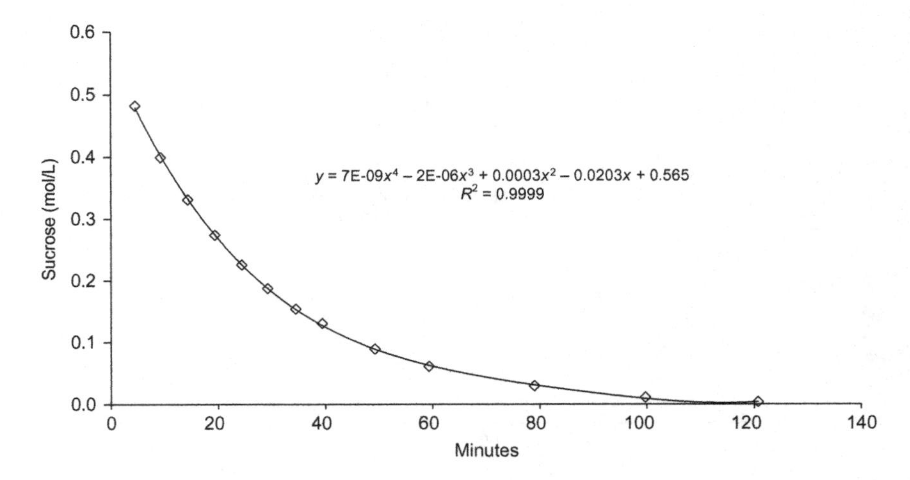

Figure 3.8 Sucrose concentration as a function of time.

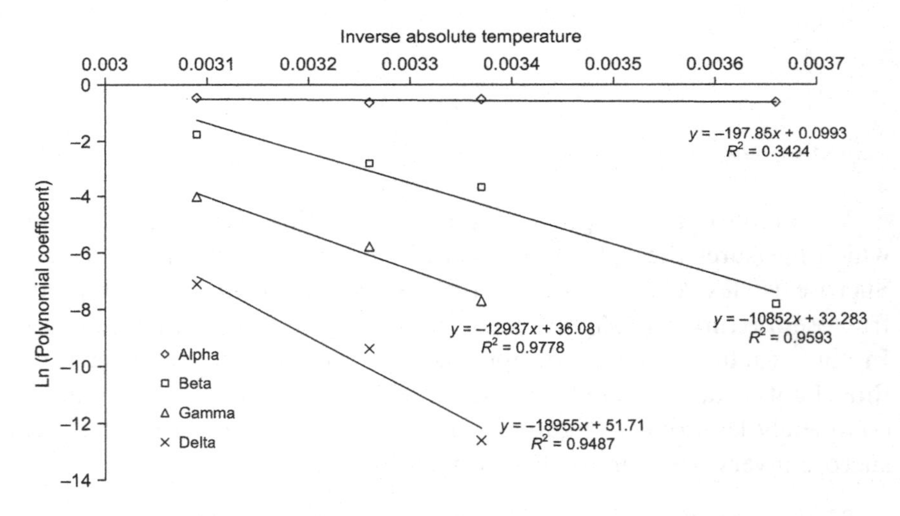

Figure 3.9 Temperature dependence of polynomial coefficients.

$$\beta = Ae^{-E/RT}$$

Similar plots and regressions can be made for sucrose inversions done at different temperatures. From such experiments, we can make an Arrhenius plot, similar to Figure 3.6, for sucrose inversion. Figure 3.9 shows such a plot.

Table 3.1 tabulates the Arrhenius slopes of each polynomial coefficient for ethyl acetate saponification and sucrose inversion. These

Table 3.1 Relationship of Polynomial Coefficient and Slope of Arrhenius Plot

Polynomial Coefficient	Slope of Arrhenius for Ethyl Acetate Saponification	Slope of Arrhenius for Sucrose Inversion
α	2839	0
β	2413	10,900
γ	3388	12,900
δ	3989	19,000
ε	5402	
ζ	3541	
ε	4108	

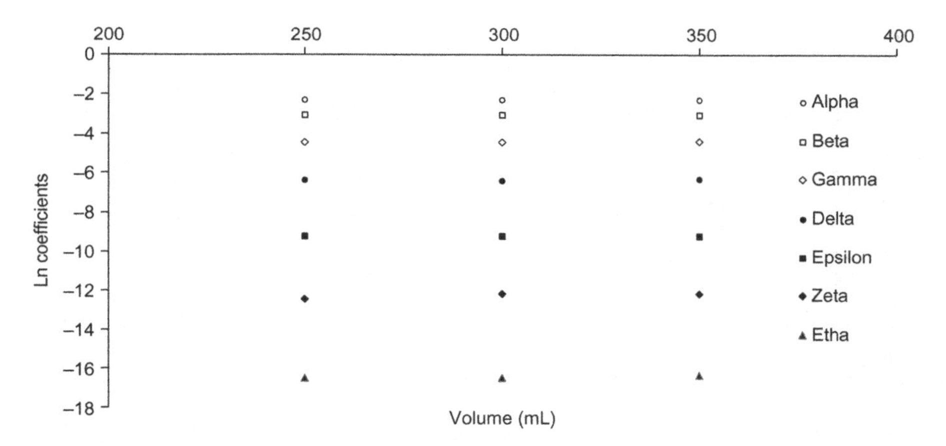

Figure 3.10 Volume dependency of polynomial coefficients of ethyl acetate saponification.

slopes are not activation energies; to convert to activation energies, we need to divide each by R, the gas constant. These slopes show that the β, γ, and δ coefficients depend upon process temperature and that they increase as $\beta < \gamma < \delta$. ζ and η also depend upon process temperature, but whether a pattern exists or not is unclear. α depends upon process temperature if the polynomial fits product formation. α is independent of process temperature if the polynomial fits reactant consumption.

Figures 3.10 and 3.11 display polynomial coefficients as a function of reactor volume. Both figures demonstrate that polynomial coefficients are independent of reactor volume, where we define reactor volume as the liquid volume within the reactor. Figure 3.10 shows the polynomial coefficients for ethyl acetate saponification to be independent of reactor volume while Figure 3.11 shows the same result for sucrose inversion.[1,3]

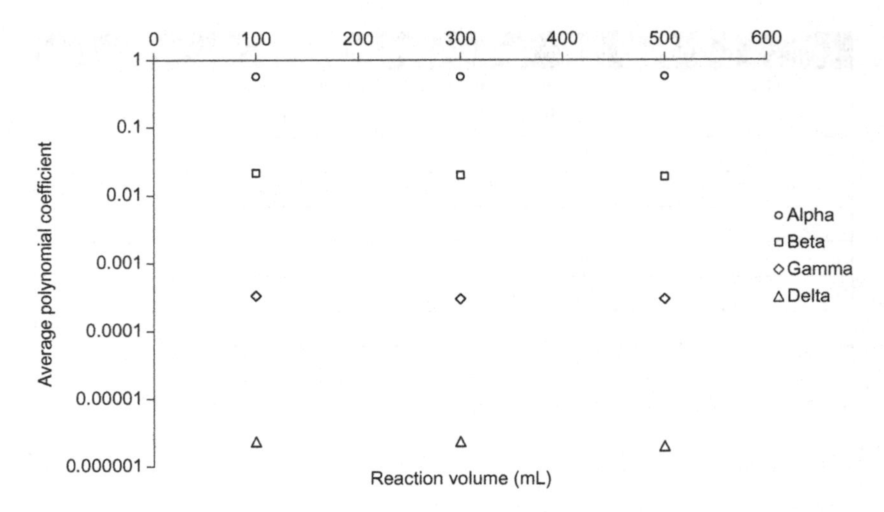

Figure 3.11 Volume dependency of polynomial coefficients for sucrose inversion.

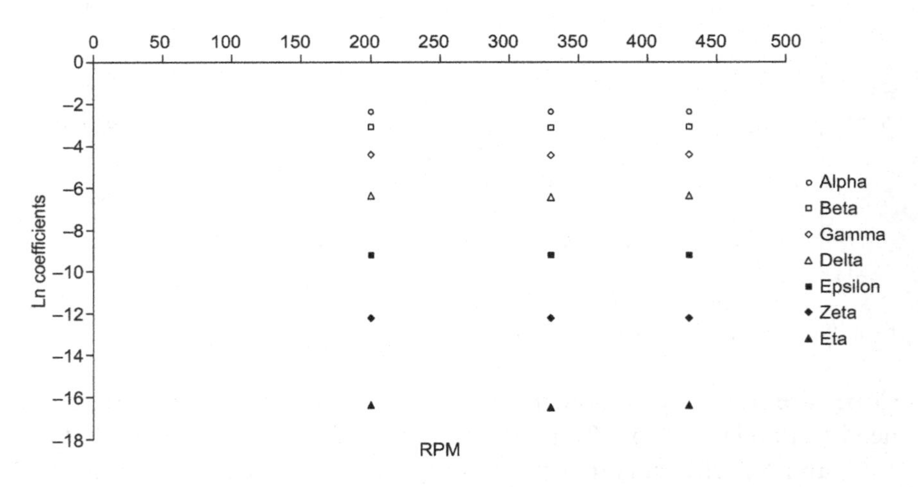

Figure 3.12 RPM dependence of polynomial coefficients for ethyl acetate saponification.

Thus, we can use a product formation curve or a reactant consumption curve for any reaction volume so long as all other reaction variables remain unchanged from the original reaction size.

When we operate a batch reactor, we inherently assume that molecular diffusion is negligible compared to the physically induced agitation of the reacting mixture. Figure 3.12 displays individual polynomial coefficients as a function of RPM (revolutions per minute), where RPM represents quality of agitation. This figure shows that the

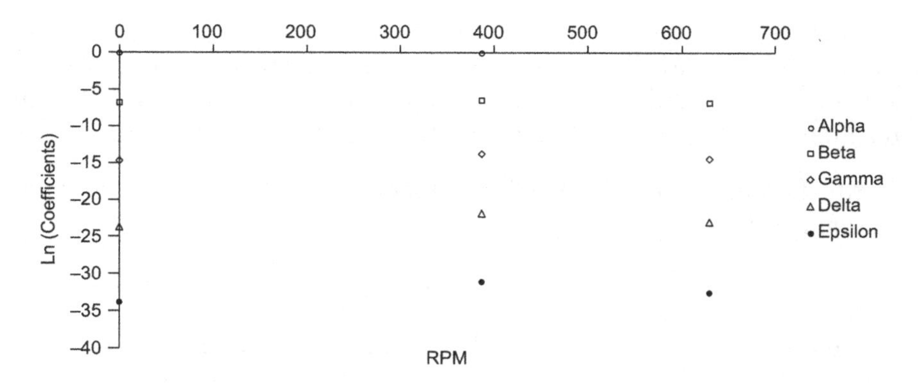

Figure 3.13 RPM dependence of polynomial coefficients for sucrose inversion. 1 M Sucrose, 65° C.

polynomial coefficients are independent of RPM for ethyl acetate saponification.[1] Figure 3.13 plots polynomial coefficient as a function of RPM for sucrose inversion.[4] Again, the polynomial coefficients are independent of RPM, i.e., of agitation.

SCALING HEAT TRANSFER

Laboratory-sized batch reactors are seldom heat transfer rate limited. Commercial-sized batch reactors are seldom not heat transfer rate limited. This conundrum arises from the energy balance for batch reactors, which, for a perfectly mixed reaction fluid, is

$$\frac{\mathrm{d}(V\rho H)}{\mathrm{d}t} = -\Delta H_{\mathrm{Rxn}} VR(C_{\mathrm{LR}}, T) + P_{\mathrm{Mix}} + U_{\mathrm{O}} A_X (T_X - T)$$

where V is the reaction volume (m^3); ρ, process fluid density ($\mathrm{kg/m}^3$); H, specific process fluid enthalpy ($\mathrm{m}^2/\mathrm{s}^2$); ΔH_{Rxn}, heat generated by the reaction (the negative sign indicates an exothermic reaction; $\mathrm{kg*m}^2/\mathrm{mol*s}^2$); $R(C_{\mathrm{LR}},T)$, reaction rate ($\mathrm{mol/m}^3\mathrm{*s}$—note, R is a function of the limiting reagent concentration C_{LR}, and process temperature T); P_{mix}, energy transferred from the process propeller or turbine mixer to the liquid within the reactor ($\mathrm{kg*m}^2/\mathrm{s}^3$); U_{O}, overall heat transfer coefficient ($\mathrm{kg/s}^3\mathrm{*K}$); A_X, external surface area available for heat transfer (m^2); T_X, heat transfer fluid temperature (K); T, process temperature (K). Thus, each term has units of power ($\mathrm{kg*m}^2/\mathrm{s}^3$).

The energy balance illuminates the problem we confront when scaling a batch reaction from the laboratory size or pilot plant size to the commercial size: the reaction volume V grows faster than the external surface area A_X available for heat transfer. In other words, at some reactor size, an isothermal batch reaction becomes an adiabatic batch reaction: at that point we lose ability to control process temperature.

Most commercial-sized batch reactors are circular cylinders with an external surface area of

$$A_X = 2\pi r^2 + 2\pi r L$$

where r is radius and L is length. The volume of such a reactor is

$$V = 2\pi r^2 L$$

Differentiating V with respect to L gives

$$\frac{dV}{dL} = 2\pi r^2$$

and differentiating A_X with respect to L yields

$$\frac{dA_X}{dL} = 2\pi r$$

Dividing the former equation by the latter equation shows that

$$\frac{dV/dL}{dA_X/dL} = \frac{dV}{dA_X} = \frac{r}{2}$$

Thus, reactor volume increases faster than reactor external surface area when scaling to commercial size. If we operate a small reactor isothermally, then increasing reactor volume moves the reactor toward adiabatic operation.

We can determine when reactor volume equals reactor external surface area. Solving for L in the equation for A_X yields

$$\frac{A_X}{2\pi r} - r = L$$

then substituting for L in the equation for V gives us

$$V = 2\pi r^2 \left(\frac{A_X}{2\pi r} - r \right) = \frac{r A_X}{2} - 2\pi r^3$$

Differentiating V above with respect to r yields

$$\frac{\mathrm{d}V}{\mathrm{d}r} = \frac{A_X}{2} - 2\pi r^2$$

Thus, the maximum achievable V at A_X is[5]

$$A_X^{\mathrm{Max}} = 4\pi r^2$$

Substituting A_X^{Max} for A_X in

$$A_X = 2\pi r^2 + 2\pi rL$$

then solving for L shows that external surface area equals volume at

$$L = r$$

Therefore, after attaining $L = r$, an isothermal process in a batch reactor rapidly becomes adiabatic.

For a constant volume, constant physical properties process, the energy balance reduces to

$$\frac{\mathrm{d}T}{\mathrm{d}t} = \frac{-\Delta H_{\mathrm{Rxn}} R(C_{\mathrm{LR}}, T)}{\rho C_{\mathrm{P}}} + \frac{P_{\mathrm{Mix}}}{\rho V C_{\mathrm{P}}} + \frac{U_{\mathrm{O}} A_X}{\rho V C_{\mathrm{P}}}(T_X - T)$$

where C_{P} is the process fluid specific heat capacity at constant pressure ($\mathrm{m}^2/\mathrm{s}^2{*}\mathrm{K}$).[6] Note that $R(C_{\mathrm{LR}},T)$ connects the energy balance to the mass balance. Thus, our need to determine $R(C_{\mathrm{LR}},T)$. For an adiabatic reaction in which P_{mix} is negligible, the energy equation for a constant volume, constant physical properties process becomes

$$\frac{\mathrm{d}T}{\mathrm{d}t} = \frac{-\Delta H_{\mathrm{Rxn}} R(C_{\mathrm{LR}}, T)}{\rho C_{\mathrm{P}}}$$

Rearranging the above equation gives

$$\mathrm{d}T = \left\{\frac{-\Delta H_{\mathrm{Rxn}} R(C_{\mathrm{LR}}, T)}{\rho C_{\mathrm{P}}}\right\}\mathrm{d}t$$

Historically, we have assumed $R(C_{\mathrm{LR}},T)$ to be k[LR], $k[\mathrm{LR}]^2$, or k_{Pseudo}[LR] or some other simple form that relates limiting reactant concentration [LR] to $R(C_{\mathrm{LR}},T)$. We made this assumption so we could integrate the above equation. If we assume the reaction is first order in limiting reactant, then the above equation becomes

$$\mathrm{d}T = \left\{ \frac{-\Delta H_{\mathrm{Rxn}}(-k[\mathrm{LR}])}{\rho C_{\mathrm{P}}} \right\} \mathrm{d}t$$

which upon integrating yields

$$T = \left\{ \frac{-\Delta H_{\mathrm{Rxn}}(-k[\mathrm{LR}])}{\rho C_{\mathrm{P}}} \right\} t + \kappa$$

where κ is an integration constant. At $t = 0$, $\kappa = T_0$; thus, the above equation becomes

$$T - T_0 = \left\{ \frac{-\Delta H_{\mathrm{Rxn}}(-k[\mathrm{LR}])}{\rho C_{\mathrm{P}}} \right\} t$$

When we declare the reaction complete, the process temperature will be T, depending upon the validity of our assumption for $R(C_{\mathrm{LR}},T)$. If our assumption for $R(C_{\mathrm{LR}},T)$ is not particularly accurate, i.e., valid, then the resulting process temperature may be significantly higher or lower than what we calculated for the reaction. If it is lower, our surprise will be an intellectual one. However, if the actual process temperature is higher than our calculated process temperature, then our surprise may involve a safety incident.

Reconsider the adiabatic energy balance

$$\mathrm{d}T = \left\{ \frac{-\Delta H_{\mathrm{Rxn}}}{\rho C_{\mathrm{P}}} \right\} R(C_{\mathrm{LR}}, T)\mathrm{d}t$$

We have shown above that a polynomial equation can be fitted to a curve relating the concentration of the limiting reactant to time, which has the general form

$$[\mathrm{LR}] = \alpha - \beta t + \gamma t^2 - \delta t^3 + \varepsilon t^4 - \zeta t^5 + \eta t^6$$

Differentiating this equation with respect to time gives

$$\frac{\mathrm{d}[\mathrm{LR}]}{\mathrm{d}t} = -\beta + 2\gamma t - 3\delta t^2 + 4\varepsilon t^3 - 5\zeta t^4 + 6\eta t^5$$

But, from the mass balance for a batch reactor

$$\frac{\mathrm{d}[\mathrm{LR}]}{\mathrm{d}t} = R(C_{\mathrm{LR}}, T)$$

Therefore

$$\frac{\mathrm{d[LR]}}{\mathrm{d}t} = R(C_{\mathrm{LR}}, T) = -\beta + 2\gamma t - 3\delta t^2 + 4\varepsilon t^3 - 5\zeta t^4 + 6\eta t^5$$

Substituting the above equation into the adiabatic energy balance gives us

$$\mathrm{d}T = \left\{\frac{-\Delta H_{\mathrm{Rxn}}}{\rho C_{\mathrm{P}}}\right\}(-\beta + 2\gamma t - 3\delta t^2 + 4\varepsilon t^3 - 5\zeta t^4 + 6\eta t^5)\mathrm{d}t$$

Integrating this equation yields

$$\int_{T_0}^{T} \mathrm{d}T = \left\{\frac{-\Delta H_{\mathrm{Rxn}}}{\rho C_{\mathrm{P}}}\right\} \int_0^t (-\beta + 2\gamma t - 3\delta t^2 + 4\varepsilon t^3 - 5\zeta t^4 + 6\eta t^5)\mathrm{d}t$$

which is

$$T - T_0 = \left\{\frac{-\Delta H_{\mathrm{Rxn}}}{\rho C_{\mathrm{P}}}\right\}(-\beta t + \gamma t^2 - \delta t^3 + \varepsilon t^4 - \zeta t^5 + \eta t^6)$$

Note that the above result does not contain an assumption; ΔH_{Rxn}, ρ, and C_{P} are experimentally measured constants. We determine the polynomial coefficients by fitting an equation to experimentally derived data. The inherent safety of a chemical process increases as we remove assumptions from its mathematical analysis.

SUMMARY

This chapter discussed various methods for scaling batch reactors. It also discussed heat transfer into and from batch reactors.

REFERENCES

1. Bor-Yea Hsu. University of Houston, Private communication. August 2014.
2. House J. *Principles of chemical kinetics*. Dubuque, IA: Wm. C. Brown Publishers; 1997. p. 31–35
3. Broussard B. University of Houston, Private communication. August 2013.
4. Ferguson C. University of Houston, Private communication. August 2014.
5. Franklin P. *Differential and integral calculus*. New York, NY: Dover Publications, Inc.; 1969. p. 60–61 (originally published by McGraw-Hill Book Company, Inc. during 1953).
6. Nauman E. *Chemical reactor design, optimization, and scaleup*. 2nd ed. New York, NY: John Wiley & Sons, Inc.; 2008. p. 172

CHAPTER 4

Semi-Batch Reactors

REACTOR CHOICE: BATCH OR SEMI-BATCH REACTOR

Consider the oxidation of an alcohol with molecular oxygen

$$R\text{–}CH_2OH + O_2 \rightarrow R\text{–}COOH + H_2O$$

This reaction is highly exothermic. If we place a complete charge of both reactants in a batch reactor, we will be unable to control the temperature of the reaction. However, if we charge the reactor with alcohol and define oxygen as the limiting reagent and meter it into the reactor at a rate controlled by the temperature of the reaction volume, then we can conduct this oxidation safely.

Intermittently or continuously feeding material to or withdrawing material from a batch reactor converts it into a semi-batch reactor. Note, we either feed material to or withdraw material from the reactor; we do not do both simultaneously. If we simultaneously feed material to and withdraw material from a reactor, then we operate the reactor as a continuous stirred tank reactor (CSTR).

Next, consider the epoxidation of an alcohol to produce a hydrophilic surfactant

$$R\text{–}CH_2OH + C_2H_4O \rightarrow R\text{–}CH_2O\text{–}C_2H_4OH$$

This reaction is catalyzed by various chemical reagents. Ethylene oxide, C_2H_4O, is a gas at ambient temperatures and is highly reactive. If we place a complete charge of it in batch reactor that contains alcohol, we will not be able to control the temperature of the reactor or control the distribution of products. In many cases, however, a product distribution is not our goal. Our goal may be to add only one ethylene oxide molecule per alcohol molecule. Controlling reaction temperature and meeting product specification require us to define ethylene oxide as the limiting reagent for this reaction and to feed it to the reactor at a predetermined rate.

Batch and Semi-batch Reactors.

Similar situations arise for two liquid reactants. If one liquid reactant is highly reactive or if the reaction is highly exothermic, we will charge a reactor completely with one reactant, then meter the second reactant into the reactor.

In general, we use a semi-batch reactor

- when the reaction is highly exothermic;
- when side reactions occur, thereby forming undesired by-products;
- when reaction occurs instantaneously upon contact of reactant;
- when equilibrium limits the reaction;
- when one reactant is sparingly soluble in a second reactant, as with gas–liquid reactions and immiscible liquid–liquid reactions;
- when the initial reactant undergoes a sequence of reactions, as in the pharmaceutical industry.[1]

We quantify by-product formation with "selectivity." There are various definitions for selectivity. One definition is based on yields: it is the overall selectivity and is the ratio of product yield to by-product yield.[2] Mathematically

$$S_{\text{Overall}} = \frac{([\text{R}]_{t=0} - [\text{R}]_t)_{\text{Prod}}/[\text{R}]_{t=0}}{\sum([\text{R}]_{t=0} - [\text{R}]_t)_{\text{ByProd}}/[\text{R}]_{t=0}} = \frac{x_{\text{Prod}}}{\sum x_{\text{ByProd}}}$$

where S_{Overall} is the overall selectivity; $[\text{R}]_{t=0}$ is the reactant concentration at time zero; $[\text{R}]_t$ is the reactant concentration at time t; x_{Prod} is the conversion of reactant to product; and Σx_{ByProd} is the summed conversion of reactant to all by-products.

Another definition of selectivity is based on the ratio of product formation rate to by-product formation rate, which is called point selectivity. It is

$$S_{\text{Point}} = \frac{\text{d}[\text{P}]/\text{d}t}{\text{d}[\text{BP}]/\text{d}t} = \frac{R_{\text{Prod}}}{R_{\text{ByProd}}}$$

where S_{Point} is point selectivity; [P] is product concentration; [BP] is by-product formation; and t is time.[3]

Average selectivity is determined at the end of a reaction. It is the amount of product in the reactor divided by the amount of reactant consumed; in other words

$$S_{\text{Average}} = \frac{n_{\text{Prod}}}{n_{\text{Reactant},t=0} - n_{\text{Reactant,End}}}$$

where $S_{Average}$ is the average selectivity; n_{Prod} is the moles of product at the end of the reaction; $n_{Reactant,t=0}$ is the moles of reactant at the start of reaction; and $n_{Reactant,End}$ is the moles of reactant at the end of the reaction.[4]

The three main mechanisms for by-product formation are

- simultaneous reactions

$$A \rightarrow P$$
$$A \rightarrow BP$$

- parallel or concurrent reactions

$$A \rightarrow P$$
$$B \rightarrow BP$$

- consecutive linear reactions

$$A \rightarrow P \rightarrow BP$$

where A represents reactants; P, product; and, BP, by-product.

Consider the most general form of the simultaneous reaction mechanism:

$$A + B \rightarrow P$$
$$A + B \rightarrow BP$$

We will assume, for this example, that, for product formation, α and β are the reaction orders for A and B, respectively, and that, for by-product formation, γ and δ are the reaction orders for A and B, respectively. The rate equations are then

$$\frac{\mathrm{dP}}{\mathrm{d}t} = k_{\mathrm{P}}[\mathrm{A}]^{\alpha}[\mathrm{B}]^{\beta} \quad \text{and} \quad \frac{\mathrm{dBP}}{\mathrm{d}t} = k_{\mathrm{BP}}[\mathrm{A}]^{\gamma}[\mathrm{B}]^{\delta}$$

Point selectivity for this generalized process is

$$S_{\mathrm{P}} = \frac{\mathrm{dP}/\mathrm{d}t}{\mathrm{dBP}/\mathrm{d}t} = \frac{k_{\mathrm{P}}[\mathrm{A}]^{\alpha}[\mathrm{B}]^{\beta}}{k_{\mathrm{BP}}[\mathrm{A}]^{\gamma}[\mathrm{B}]^{\delta}}$$

If $\alpha = \beta$ and $\gamma = \delta$, then

$$S_{\mathrm{P}} = \frac{k_{\mathrm{P}}}{k_{\mathrm{BP}}}$$

Since S_P is independent of all reactant concentrations, we would choose a batch reactor for this process, provided we can control the temperature of the reaction. If $\alpha > \gamma$ and $\beta = \delta$, then

$$S_P = \frac{k_P[A]^\alpha[B]^\beta}{k_{BP}[A]^\gamma[B]^\delta} = \frac{k_P[A]^{\alpha-\gamma}}{k_{BP}}$$

Thus, high [A] favors P formation while high [B] maximizes the rate of P formation. In this case, we would, again, choose a batch reactor, provided we can control the temperature of the reaction. If $\alpha < \gamma$ and $\beta = \delta$, then

$$S_P = \frac{k_P}{k_{BP}[A]^{\gamma-\alpha}}$$

In this case, we maximize S_P by maintaining a low [A]. Maintaining a high [B] ensures a high P formation rate. Thus, we would choose a semi-batch reactor for this process: we would charge the reactor with B and feed A at a rate appropriate to obtain a high S_P. If $\alpha < \gamma$ and $\beta < \delta$, then

$$S_P = \frac{k_P}{k_{BP}[A]^{\gamma-\alpha}[B]^{\delta-\beta}}$$

which indicates that we must keep both [A] and [B] low to maintain a high S_P. In this case, we would not use a batch or a semi-batch reactor for our process. Rather, we would use a single CSTR for this process. If $\alpha > \gamma$ and $\beta > \delta$, then

$$S_P = \frac{k_P[A]^\alpha[B]^\beta}{k_{BP}[A]^\gamma[B]^\delta} = \left(\frac{k_P}{k_{BP}}\right)[A]^{\alpha-\gamma}[B]^{\beta-\gamma}$$

which indicates that both [A] and [B] must be maintained as high as possible to maximize S_P. In this case, we would choose a plug flow reactor rather than a batch or semi-batch reactor.[5(pp276–277)]

Examples of simultaneous reactions are ethylene oxidation to produce ethylene oxide[6]

$$H_2C{=}CH_2 + (1/2)O_2 \rightarrow EO$$
$$H_2C{=}CH_2 + 3O_2 \rightarrow 2CO_2 \rightarrow 2H_2O$$

and hydrocarbon carbon, which, for simplicity, we present as propane cracking[7]

$$C_3H_8 \rightarrow C_3H_6 + H_2$$
$$C_3H_8 \rightarrow C_2H_6 + CH_4$$

The general representation for concurrent or parallel reactions is

$$\begin{gathered}A + B \rightarrow P \\ C + D \rightarrow BP\end{gathered}$$

where

$$\frac{dP}{dt} = k_P[A][B] \quad \text{and} \quad \frac{dBP}{dt} = k_{BP}[C][D]$$

In this case, we do not worry about the order of reaction for each reactant since the reactants are independent of each other. Point selectivity for this generalized process is

$$S_P = \frac{dP/dt}{dBP/dt} = \frac{k_P[A][B]}{k_{BP}[C][D]}$$

If C enters the process with A and D enters the process with B, then we would use a semi-batch reactor. Which reactant is completely charged to the reactor and which reactant is fed to the reactor during the process is a matter of our convenience. If C and D enter the process with either A or B, then we would, again, use a semi-batch reactor; however, we would charge the reactor with the C- and D-free reactant and feed the other reactant to it.

The last simple, classic reaction mechanism is conservative linear reactions. Its generalized form is

$$\begin{gathered}A + B \rightarrow P \\ P + B \rightarrow BP\end{gathered}$$

Conservative linear mechanisms are common in the CPI. They include

- benzene chlorination;
- multiple olefin hydrogenation;
- ethylene oxide hydrolysis;
- hydrocarbon partial oxidation.[4(pp85–86)]

If we assume first-order rate dependency for all reactant concentrations, then

$$\frac{dP}{dt} = k_P[A][B] - k_{BP}[P][B] \quad \text{and} \quad \frac{dBP}{dt} = k_{BP}[P][B]$$

Point selectivity is then

$$S_P = \frac{dP/dt}{dBP/dt} = \frac{k_P[A][B] - k_{BP}[P][B]}{k_{BP}[P][B]} = \frac{k_P[A][B]}{k_{BP}[P][B]} - 1 = \frac{k_P[A]}{k_{BP}[P]} - 1$$

which indicates a maximum S_P when [P] is minimized. We would certainly not recommend using a batch reactor in this case: as P accumulates in the batch reactor, S_P declines. We would prefer not to use a semi-batch reactor because at some P concentration, S_P maximizes. Beyond that [P], S_P declines. Thus, we would need to stop the reaction when [P] is maximized, discharge the reactor, then separate P from A, B, and any BP present in the final brew. If we operate this separation batch-wise, it will be expensive. Our choice might be a series of CSTRs followed by continuous distillation with recycle of A and B to the reactor—provided distillation is the appropriate separation process.

Thus, knowledge of the by-products formed during a reaction and the kinetics of product formation and by-product formation are imperative when scaling a process or reaction from laboratory-sized equipment to pilot plant- or commercial-sized equipment.[8]

MISCIBLE LIQUID–LIQUID SEMI-BATCH PROCESSES

Kinetic Analysis

Consider the generalized reaction between two completely miscible liquids of similar density. Complete miscibility means no second phase, i.e., the process is homogeneous. We also assume the product is miscible in the two reactants and has a density similar to those of the reactants—at least similar enough for us to assume no density change during the process. The stoichiometry of our generalized reaction is

$$LR + BR \rightarrow P$$

where LR designates the "limiting reactant"; BR designates the "bulk reactant"; and P, the product formed by the reaction. We charge the reactor with BR before initiating the reaction. We add LR continuously to the reactor after initiating the reaction. For the process to be reaction rate limited, molecular diffusion within the reaction mixture must be negligible. Thus, we assume ideal, perfect mixing of the reacting fluid.

The component balance for the bulk reactant, therefore, reduces to

$$\frac{d(V_R[BR])}{dt} = V_R R_{BR}$$

where V_R is the volume of the reactor at time t; [BR] is the concentration of BR at time t; and R_{BR} is the rate of the chemical reaction consuming BR. Note that V_R is the liquid volume in the vessel. We use a molar balance in this case because V_R depends upon time, i.e., upon LR addition to the semi-batch reactor.

With regard to the limiting reactant, its component balance is

$$\frac{d(V_R[LR])}{dt} = [LR]_{Feed} Q_{Feed} + V_R R_{LR}$$

where [LR] is the concentration of LR in V_R; Q_{Feed} is the volumetric flow rate of LR into the vessel; and R_{LR} is the chemical reaction consuming LR. Again, we assumed molecular diffusion within the reacting fluid to be negligible.

The component balance for product P is

$$\frac{d(V_R[P])}{dt} = V_R R_P$$

where [P] is the concentration of P in V_R and R_P is the formation of P in V_R.

First, all the component equations have V_R as a variable. For a constant density process, V_R increases as limiting reactant enters the reactor. In other words

$$\frac{dV_R}{dt} = Q_{Feed}$$

Rearranging and integrating gives us

$$\int_{V_0}^{V_R} dV_R = Q_{Feed} \int_0^t dt$$

$$V_R - V_0 = Q_{Feed} t$$

Thus, the reactor fluid volume at time t is

$$V_R = V_0 + Q_{Feed} t$$

Second, each of the component equations depends upon a consumption rate expression or a generation rate expression. It is reasonable to assume that each of these rate expressions depends upon reactant concentration in some fashion. We assume, for this analysis, that each rate expression is first order with respect to each reactant; mathematically they are

$$R_{BR} = -k[LR][BR]$$
$$R_{LR} = -k[LR][BR]$$
$$R_{P} = k[LR][BR]$$

Remember, the negative sign indicates consumption while a positive sign indicates formation. We can now write the component mass balance for the bulk reactant as

$$\frac{d(V_R[BR])}{dt} = -k[LR][BR]V_R$$

Expanding the differential term yields

$$V_R\frac{d[BR]}{dt} + [BR]\frac{dV_R}{dt} = -k[LR][BR]V_R$$

Rearranging gives

$$V_R\frac{d[BR]}{dt} = -k[LR][BR]V_R - [BR]\frac{dV_R}{dt}$$

But, from above

$$\frac{dV_R}{dt} = Q_{Feed}$$

Thus

$$V_R\frac{d[BR]}{dt} = -k[LR][BR]V_R - Q_{Feed}[BR]$$

Dividing by V_R yields

$$\frac{d[BR]}{dt} = -k[LR][BR] - \frac{Q_{Feed}}{V_R}[BR]$$

Doing the same for the other two component balances gives us

$$\frac{d[LR]}{dt} = \frac{Q_{Feed}}{V_R}[LR]_{Feed} - \frac{Q_{Feed}}{V_R}[LR] - k[LR][BR]$$

and

$$\frac{d[P]}{dt} = k[LR][BR] - \frac{Q_{Feed}}{V_R}[P]$$

These three differential equations can be solved by a variety of software packages.[9] However, by making judicious assumptions about the semi-batch process, we can solve the above rate equations analytically. First, operationally, we completely charge the bulk reactant to the reactor and slowly add the limiting reactant to it, either to control the rate of heat release or to control by-product formation. Therefore, at any given time, the moles of bulk reactant exceeds, and sometimes greatly exceeds, the moles of limiting reactant. In such a situation, we can assume a pseudo-first-order reaction with respect to the limiting reactant.[10] In other words

$$R = k[\mathrm{LR}][\mathrm{BR}] = k_{\mathrm{Pseudo}}[\mathrm{LR}]$$

where

$$k_{\mathrm{Pseudo}} = k[\mathrm{BR}]$$

This assumption holds true so long as $[\mathrm{BR}] \gg [\mathrm{LR}]$. We must validate this assumption upon concluding our mathematical analysis of these rate equations. Upon utilizing this assumption, the above rate equations become

$$\frac{\mathrm{d}[\mathrm{BR}]}{\mathrm{d}t} = -k_{\mathrm{Pseudo}}[\mathrm{LR}] - \frac{Q_{\mathrm{Feed}}}{V_{\mathrm{R}}}[\mathrm{BR}]$$

$$\frac{\mathrm{d}[\mathrm{P}]}{\mathrm{d}t} = k_{\mathrm{Pseudo}}[\mathrm{LR}] - \frac{Q_{\mathrm{Feed}}}{V_{\mathrm{R}}}[\mathrm{P}]$$

$$\frac{\mathrm{d}[\mathrm{LR}]}{\mathrm{d}t} = \frac{Q_{\mathrm{Feed}}}{V_{\mathrm{R}}}[\mathrm{LR}]_{\mathrm{Feed}} - \frac{Q_{\mathrm{Feed}}}{V_{\mathrm{R}}}[\mathrm{LR}] - k_{\mathrm{Pseudo}}[\mathrm{LR}]$$

Note that $\mathrm{d}[\mathrm{BR}]/\mathrm{d}t$ and $\mathrm{d}[\mathrm{P}]/\mathrm{d}t$ both depend upon [LR]; also, note that $\mathrm{d}[\mathrm{LR}]/\mathrm{d}t$ only depends upon [LR].

Second, physically, we know that the concentration of limiting reactant is low and is constant in the reacting fluid. Since we maintain a constant [LR] in the reacting fluid, $\mathrm{d}[\mathrm{LR}]/\mathrm{d}t = 0$. Thus

$$\frac{\mathrm{d}[\mathrm{LR}]}{\mathrm{d}t} = \frac{Q_{\mathrm{Feed}}}{V_{\mathrm{R}}}[\mathrm{LR}]_{\mathrm{Feed}} - \frac{Q_{\mathrm{Feed}}}{V_{\mathrm{R}}}[\mathrm{LR}] - k_{\mathrm{Pseudo}}[\mathrm{LR}] = 0$$

Solving for [LR], we obtain

$$[\mathrm{LR}] = \frac{(Q_{\mathrm{Feed}}/V_{\mathrm{R}})[\mathrm{LR}]_{\mathrm{Feed}}}{(Q_{\mathrm{Feed}}/V_{\mathrm{R}}) + k_{\mathrm{Pseudo}}}$$

Table 4.1 Operating Parameters for the Generic Chemical Reaction $BR + LR \rightarrow P$ where BR is Bulk Reactant; LR is Limiting Reactant; and, P is Product	
Operating Condition #1	
$V_0 = 100$ L $Q_{Feed} = 500$ L/h	$[BR]_{Feed} = 10$ moles/L $[LR]_{Feed} = 5$ moles/L $k = 100$ 1/h
Operating Condition #2	
$V_0 = 100$ L $Q_{Feed} = 500$ L/h	$[BR]_{Feed} = 10$ moles/L $[LR]_{Feed} = 7.1$ moles/L $k = 100$ 1/h

Substituting [LR] into the equation for d[BR]/dt, integrating, and solving for [BR] gives us

$$[BR] = \frac{V_R}{Q_{Feed}}\left(\left(\frac{(Q_{Feed}/V_R)[LR]_{Feed}}{(Q_{Feed}/V_R) + k_{Pseudo}}\right)k_{Pseudo} + \frac{Q_{Feed}}{V_R}[BR]_{Feed}e^{-(Q_{Feed}/V_R)t}\right) - \left(\frac{V_R}{Q_{Feed}}\right)\left(\frac{(Q_{Feed}/V_R)[LR]_{Feed}}{(Q_{Feed}/V_R) + k_{Pseudo}}\right)k_{Pseudo}$$

Making the same substitution for d[P]/dt, then solving for [P] yields

$$[P] = \left(\frac{V_R}{Q_{Feed}}\right)k_{Pseudo}[LR](1 - e^{-(Q_{Feed}/V_R)t})$$

$$[P] = \frac{k_{Pseudo}[LR]_{Feed}}{(Q_{Feed}/V_R) + k_{Pseudo}}(1 - e^{-(Q_{Feed}/V_R)t})$$

We can use Excel to solve these equations for [BR], [LR], and [P].

Table 4.1 contains two sets of operating conditions for a semi-batch reactor conducting the generic reaction

$$BR + LR \rightarrow P$$

where BR, LR, and P have the same definitions as above. Figure 4.1 displays the results of an Excel calculation using Operating Condition #1 shown in Table 4.1. Figure 4.1 shows that [LR] remains low throughout the reaction period and, while not strictly at steady state, is not changing greatly as the reaction progresses. Figure 4.1 also shows that [BR] decreases to a plateau of 1.2–1.5 mol/L while [P] increases to 2.8 mol/L. Thus, a subsequent separation is required to obtain pure product P. Note that [BR]/[LR] starts at 36, drops to 33, then increases to 42 during

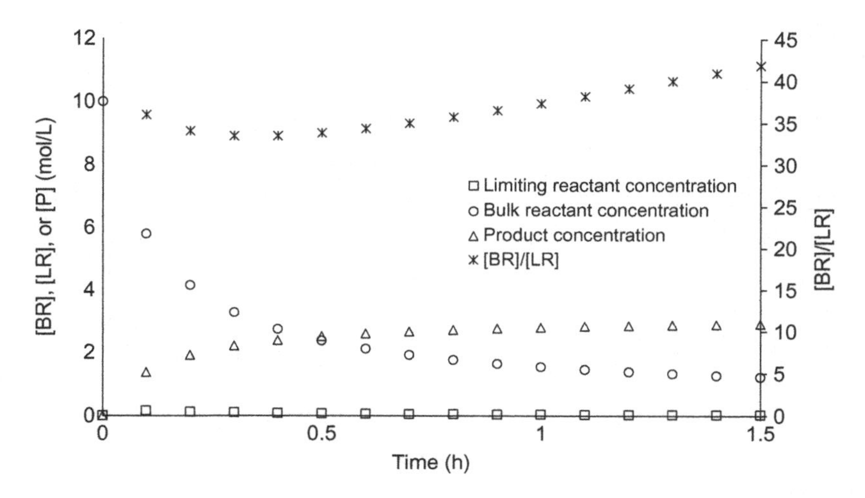

Figure 4.1 Semi-batch reactor analysis for the generic reaction, $BR + LR \rightarrow P$.

the reaction period shown in Figure 4.1, further suggesting that $d[LR]/dt$ is not absolutely zero. However, our assumption of a pseudo-first-order reaction mechanism is valid throughout the reaction period.

The reason to build models of reacting processes is to increase laboratory safety and to reduce the number of experiments required in the laboratory and pilot plant. If we do not want to separate BR and LR from P, then we must determine the operating conditions which yield that result. We can determine those operating conditions via experimentation, which is time consuming and involves some risk of mishap, or we can determine the necessary operating conditions via calculation using a process model.

Figure 4.2 displays the results for Operating Condition #2 shown in Table 4.1. It shows that [BR] is zero after 1.5 h of reaction. [LR] is 0.04 mol/L, which may or may not require subsequent separation, and [P] is 4 mol/L at 1.5 h. Note, however, that [BR]/[LR] falls below 10 after 0.8 h of reaction. We do not know whether this last point is important or not until we perform the experiment and compare its results to those of the results of the calculation. But, in any case, this calculation has saved considerable time and effort in the laboratory already, which means it has increased the productivity of our organization.

Thermal Analysis

The thermal balance for this constant density semi-batch process operating at constant temperature is

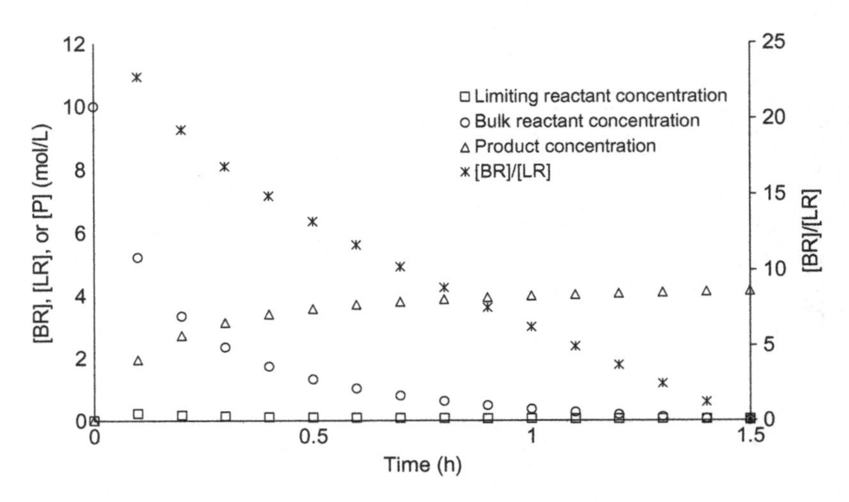

Figure 4.2 Semi-batch reactor analysis for the generic reaction, BR + LR → P.

$$(n_{BR}c_{P/BR} + n_{LR}c_{P/LR} + n_P c_{P/P})\frac{dT}{dt} = \Delta H_{BR} R_{BR}(V_0 + Q_{Feed}t)$$

$$+ Q_{Feed}[LR]c_{P/LR}(T_{Feed} - T) - U\left(\frac{S_{HeatTransfer}}{V_{Liquid}}\right)(V_0 + Q_{Feed}t)(T - T_{Jacket})$$

where n_{BR}, n_{LR}, and n_P are the respective moles of BR, LR, and P; $c_{P/BR}$, $c_{P/LR}$, and $c_{P/P}$ are the respective heat capacities of BR, LR, and P ($m^2kg/mol * K * s^2$); T is the process temperature (K); t is the time (s); $(\Delta H_{BR})_T$ is the heat of reaction for BR at temperature T ($m^2kg/mol * s^2$); R_{BR} is the kinetic reaction rate of BR (mol/m^3*s); V_0 is the initial volume of BR in the reactor (m^3); Q_{Feed} is the LR volumetric flow rate (m^3/s); [LR] is the feed concentration (mol/m^3); T_{Feed} is the feed temperature of LR (K); U is the overall heat transfer coefficient ($kg/K*s^3$); $S_{HeatTransfer}/V_{Liquid}$ is the heat transfer surface area to liquid volume (m^2/m^3); and T_{Jacket} is the temperature (K) of the heat transfer fluid flowing through the reactor jacket. The units for each of the above terms are m^2*kg/s^3, which are the units for power. When integrated, the unit for each term becomes m^2*kg/s^2, i.e., the unit of energy or work. Note that we neglected the energy added to the process by agitation, and we neglected the work done by the process as its fluid volume expands at constant pressure against the inert gas in the process headspace.[5(pp560–563)]

The above thermal balance demonstrates why semi-batch processes are popular: if we operate the process at constant temperature, then the above equation becomes

$$0 = \Delta H_{BR} R_{BR}(V_0 + Q_{Feed}t) + Q_{Feed}[LR]c_{P/LR}(T_{Feed} - T) - U\left(\frac{S_{HeatTransfer}}{V_{Liquid}}\right)(V_0 + Q_{Feed}t)(T - T_{Jacket})$$

which becomes upon rearranging

$$\Delta H_{BR} R_{BR}(V_0 + Q_{Feed}t) + Q_{Feed}[LR]c_{P/LR}(T_{Feed} - T) = U\left(\frac{S_{HeatTransfer}}{V_{Liquid}}\right)(V_0 + Q_{Feed}t)(T - T_{Jacket})$$

Thus, we control process temperature T by controlling the volumetric flow rate of LR into the process.

If our semi-batch process is exothermic and we lose flow of heat transfer fluid to the reactor jacket, the ensuing event is unlikely to become a "runaway" with regard to temperature if we block flow of LR into the process. Upon losing cooling to the process, the thermal balance for our semi-batch reactor becomes

$$(n_{BR}c_{P/BR} + n_{LR}c_{P/LR} + n_P c_{P/P})\frac{dT}{dt} = \Delta H_{BR} R_{BR}(V_0 + Q_{Feed}t) + Q_{Feed}[LR]c_{P/LR}(T_{Feed} - T)$$

when we block LR flow to the process, the above equation reduces to

$$(n_{BR}c_{P/BR} + n_{LR}c_{P/LR} + n_P c_{P/P})\frac{dT}{dt} = \Delta H_{BR} R_{BR} V_0$$

Rearranging the above equation yields

$$\frac{dT}{dt} = \frac{\Delta H_{BR} R_{BR} V_0}{(n_{BR}c_{P/BR} + n_{LR}c_{P/LR} + n_P c_{P/P})}$$

which is the adiabatic temperature rise for a batch reactor.

IMMISCIBLE LIQUID–LIQUID SEMI-BATCH PROCESSES

Kinetic Analysis

Figure 4.3 represents a semi-batch reactor process involving immiscible liquids. One liquid is completely charged to the reactor: it is the bulk reactant BR. The second liquid "sprays" into the reactor just above the surface of the bulk reactant. We call this second liquid the limiting reactant LR. LR forms micelles upon contacting BR. The agitator distributes these micelles throughout BR. The baffles assist this distribution by preventing formation of a vortex in BR.

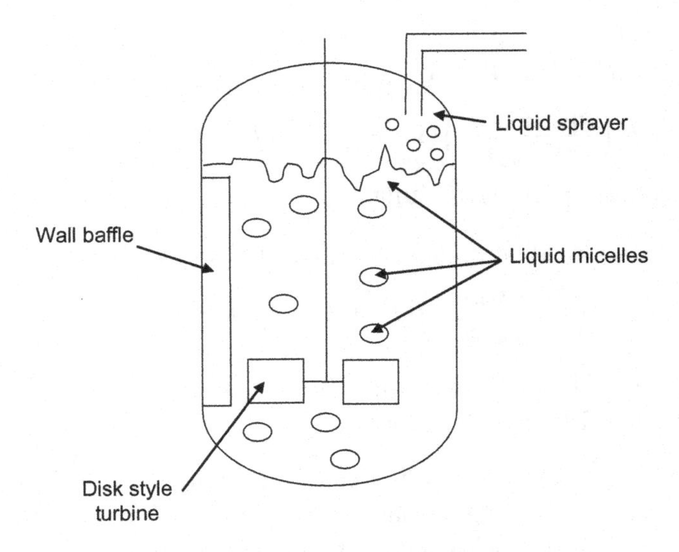

Figure 4.3 Macroscopic features of an immiscible liquids semi-batch process.

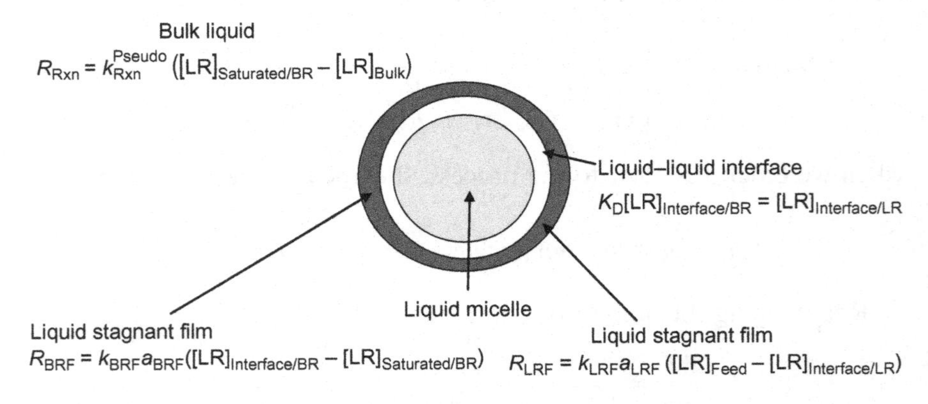

Figure 4.4 Microscopic features of gas–liquid or immiscible liquids semi-batch process.

Figure 4.4 shows the microscopic features of this process. Each LR micelle contains pure LR liquid. If LR solubilizes small amounts of BR, then a stagnant film of LR molecules exists adjacent to the liquid–liquid interface. If LR is pure and if it does not solubilize BR, then no stagnant film forms inside the LR micelle. For this analysis, we assume a stagnant film does form inside the LR micelle.[4(chp 7)] Diffusion governs the movement of liquid LR molecules within this stagnant film. Physical equilibrium, described by a distribution coefficient, controls the movement of LR across this interface and into BR.

A stagnant film of BR surrounds each LR micelle and diffusion governs the movement of solubilized LR within it. BR comprises the bulk fluid phase moving between each LR micelle.

The rate of LR movement across the stagnant film of LR is mathematically described as

$$R_{\mathrm{LRF}} = k_{\mathrm{LRF}} a_{\mathrm{LRF}} ([\mathrm{LR}]_{\mathrm{Feed}} - [\mathrm{LR}]_{\mathrm{Interface/LR}})$$

where R_{LRF} is the rate of LR molecules moving across the stagnant film of LR inside the micelle (mol/m^{3}*s); k_{LRF} is the rate constant for the movement of LR molecules across the stagnant film of LR (m/s); a_{LRF} is the surface area to volume of the stagnant film of LR inside the LR micelle (m^{2}/m^{3}); $[\mathrm{LR}]_{\mathrm{Feed}}$ is the concentration of LR entering the semi-batch reactor (mol/m^{3}); and $[\mathrm{LR}]_{\mathrm{Interface/LR}}$ is the concentration of LR at the liquid–liquid interface of the micelle (mol/m^{3}).

The distribution coefficient K_{D} is the concentration of LR in LR at the liquid–liquid interface and the concentration of LR in BR at the liquid–liquid interface;[11,12] thus, mathematically, it is

$$K_{\mathrm{D}} = \frac{[\mathrm{LR}]_{\mathrm{Interface/LR}}}{[\mathrm{LR}]_{\mathrm{Interface/BR}}}$$

where $[\mathrm{LR}]_{\mathrm{Interface/BR}}$ is the concentration of LR in BR at the liquid–liquid interface of the micelle (mol/m^{3}). Rearranging the above equation gives

$$K_{\mathrm{D}}[\mathrm{LR}]_{\mathrm{Interface/BR}} = [\mathrm{LR}]_{\mathrm{Interface/LR}}$$

The rate of movement of dissolved LR across the stagnant film of BR surrounding each micelle is

$$R_{\mathrm{BRF}} = k_{\mathrm{BRF}} a_{\mathrm{BRF}} ([\mathrm{LR}]_{\mathrm{Interface/BR}} - [\mathrm{LR}]_{\mathrm{Saturated/BR}})$$

where R_{BRF} is the rate of LR molecules moving across the stagnant film of BR surrounding the micelle (mol/m^{3}*s); k_{BRF} is the rate constant for the movement of LR molecules across the stagnant film of BR (m/s); a_{BRF} is the surface area to volume of the stagnant film of BR outside the LR micelle (m^{2}/m^{3}); $[\mathrm{LR}]_{\mathrm{Saturated/BR}}$ is the saturated concentration of LR in bulk BR (mol/m^{3}).

If the chemical reaction is irreversible, the chemical reaction within bulk BR is

$$R_{Rxn} = k_{Rxn}^{Pseudo}([LR]_{Saturated/BR} - [LR]_{Bulk})$$

We write R_{Rxn} as a first-order reaction with respect to LR because, for this process

$$[BR] \gg [LR]$$

Therefore

$$k_{Rxn}^{Pseudo} = k_{Rxn}[BR]$$

Unfortunately, in the above rate equations, we only know $[LR]_{Feed}$ and $[LR]_{Bulk}$, which we measure by proper sampling techniques and appropriate analytical methods. We must, then, reduce the above set of rate equations to one equation containing $[LR]_{Feed}$ and $[LR]_{Bulk}$.

Substituting the distribution coefficient into the first rate equation gives

$$R_{LRF} = k_{LRF}a_{LRF}([LR]_{Feed} - K_D[LR]_{Interface/BR})$$

then solving for $[LR]_{Interface/BR}$ yields

$$\frac{R_{LRF}}{k_{LRF}a_{LRF}} - [LR]_{Feed} = -K_D[LR]_{Interface/BR}$$

or

$$\frac{R_{LRF}}{K_D k_{LRF}a_{LRF}} - \frac{[LR]_{Feed}}{K_D} = -[LR]_{Interface/BR}$$

Substituting the above equation into

$$R_{BRF} = k_{BRF}a_{BRF}([LR]_{Interface/BR} - [LR]_{Saturated/BR})$$

gives

$$R_{BRF} = k_{BRF}a_{BRF}\left(-\frac{R_{LRF}}{K_D k_{LRF}a_{LRF}} + \frac{[LR]_{Feed}}{K_D} - [LR]_{Saturated/BR}\right)$$

Solving for $[LR]_{Saturated/BR}$ yields

$$\frac{R_{BRF}}{k_{BRF}a_{BRF}} + \frac{R_{LRF}}{K_D k_{LRF}a_{LRF}} - \frac{[LR]_{Feed}}{K_D} = -[LR]_{Saturated/BR}$$

Substituting this last equation into

$$R_{\mathrm{Rxn}} = k_{\mathrm{Rxn}}^{\mathrm{Pseudo}}([\mathrm{LR}]_{\mathrm{Saturated/BR}} - [\mathrm{LR}]_{\mathrm{Bulk}})$$

provides, after rearranging terms

$$\frac{R_{\mathrm{LRF}}/K_{\mathrm{D}}}{k_{\mathrm{LRF}}a_{\mathrm{LRF}}} + \frac{R_{\mathrm{BRF}}}{k_{\mathrm{BRF}}a_{\mathrm{BRF}}} + \frac{R_{\mathrm{Rxn}}}{k_{\mathrm{Rxn}}^{\mathrm{Pseudo}}} = \left(\frac{[\mathrm{LR}]_{\mathrm{Feed}}}{K_{\mathrm{D}}} - [\mathrm{LR}]_{\mathrm{Bulk}}\right)$$

We now have all the rates related to the two variables we can measure. We know the overall rate of the process is

$$R_{\mathrm{Overall}} = k_{\mathrm{Overall}}\left(\frac{[\mathrm{LR}]_{\mathrm{Feed}}}{K_{\mathrm{D}}} - [\mathrm{LR}]_{\mathrm{Bulk}}\right)$$

where R_{Overall} is the overall rate for the semi-batch process (mol/m^{3}*s) and k_{Overall} is the overall rate constant for the semi-batch process (1/s). Since none of the rates can be faster than the slowest rate and the overall rate equals the slowest rate, we can write

$$R_{\mathrm{LRF}}/K_{\mathrm{D}} = R_{\mathrm{BRF}} = R_{\mathrm{Rxn}} = R_{\mathrm{Overall}}$$

Therefore,

$$\frac{R_{\mathrm{Overall}}}{k_{\mathrm{LRF}}a_{\mathrm{LRF}}} + \frac{R_{\mathrm{Overall}}}{k_{\mathrm{BRF}}a_{\mathrm{BRF}}} + \frac{R_{\mathrm{Overall}}}{k_{\mathrm{Rxn}}^{\mathrm{Pseudo}}} = \left(\frac{[\mathrm{LR}]_{\mathrm{Feed}}}{K_{\mathrm{D}}} - [\mathrm{LR}]_{\mathrm{Bulk}}\right)$$

or

$$R_{\mathrm{Overall}}\left(\frac{1}{k_{\mathrm{LRF}}a_{\mathrm{LRF}}} + \frac{1}{k_{\mathrm{BRF}}a_{\mathrm{BRF}}} + \frac{1}{k_{\mathrm{Rxn}}^{\mathrm{Pseudo}}}\right) = \left(\frac{[\mathrm{LR}]_{\mathrm{Feed}}}{K_{\mathrm{D}}} - [\mathrm{LR}]_{\mathrm{Bulk}}\right)$$

Rearranging gives

$$R_{\mathrm{Overall}} = \left(\frac{1}{(1/k_{\mathrm{LRF}}a_{\mathrm{LRF}}) + (1/k_{\mathrm{BRF}}a_{\mathrm{BRF}}) + (1/k_{\mathrm{Rxn}}^{\mathrm{Pseudo}})}\right)\left(\frac{[\mathrm{LR}]_{\mathrm{Feed}}}{K_{\mathrm{D}}} - [\mathrm{LR}]_{\mathrm{Bulk}}\right)$$

Thus

$$k_{\mathrm{Overall}} = \left(\frac{1}{(1/k_{\mathrm{LRF}}a_{\mathrm{LRF}}) + (1/k_{\mathrm{BRF}}a_{\mathrm{BRF}}) + (1/k_{\mathrm{Rxn}}^{\mathrm{Pseudo}})}\right)$$

which becomes

$$\frac{1}{k_{\text{Overall}}} = \frac{1}{k_{\text{LRF}}a_{\text{LRF}}} + \frac{1}{k_{\text{BRF}}a_{\text{BRF}}} + \frac{1}{k_{\text{Rxn}}^{\text{Pseudo}}}$$

We generally interpret the above equation as representing a series of resistances. The overall resistance, $1/k_{\text{Overall}}$, equals the sum of the resistance to LR diffusion across the stagnant liquid film inside the micelle, the resistance to LR diffusion across the liquid film surrounding the micelle, and the resistance to LR reaction. Note, $k_{\text{Rxn}}^{\text{Pseudo}}$, k_{BRF}, and k_{LRF} are constants. $k_{\text{Rxn}}^{\text{Pseudo}}$ describes the time occurrence of the chemical reaction; that is, under a given set of operating conditions, e.g., temperature and pressure, $k_{\text{Rxn}}^{\text{Pseudo}}$ does not change. k_{BRF} and k_{LRF} depend upon the mean free path of a molecule in the liquid and the mean free path depends upon liquid density, which, in our case, we assume is constant. Therefore, any fluctuations in k_{Overall} must arise from fluctuations in a_{BRF} or a_{LRF} or both.

If the micelles of a given liquid–liquid semi-batch process are spherical, then

$$a_{\text{LRF}} \text{ or } a_{\text{BRF}} = \frac{4\pi r^2}{(4/3)\pi r^3} = \frac{3}{r}$$

where r is the radius of the micelle. Stagnant film thickness δ is a function of r; therefore,

$$a_{\text{LRF}} \text{ or } a_{\text{BRF}} \propto \frac{1}{\delta}$$

So, as the micelle shrinks due to solubilization and chemical reaction, a_{BRF} and a_{LRF} increase. An increase in a_{BRF} and a_{LRF} leads directly to an increase in k_{Overall} since

$$\frac{1}{k_{\text{Overall}}} \propto \frac{1}{a_{\text{LRF}}} \quad \text{and} \quad \frac{1}{a_{\text{BRF}}}$$

k_{Overall} increases because, as the LR and BR stagnant film thicknesses decrease, the time to cross each of them decreases due to k_{BRF} and k_{LRF} being constant.

a_{BRF} depends not only on micelle radius, but also on the velocity of BR streaming past an LR micelle. For a constant micelle radius, we obtain the thinnest BR stagnant film at the highest possible agitation rate. To further reduce BR stagnant film thickness, we must decrease the radius of the LR micelle. If we operate a semi-batch process at an agitation rate that minimizes the BR stagnant film thickness with respect to BR fluid velocity, then we can represent the process schematically as shown in Figure 4.5, where we plot $1/k_{\text{Overall}}$ as a function of micelle radius r. Since resistance to LR chemical reaction, i.e., $1/k_{\text{Rxn}}^{\text{Pseudo}}$, is independent of micelle radius, it plots as a horizontal line in Figure 4.5. Resistance to LR diffusion across the BR stagnant film surrounding an LR micelle plots in Figure 4.5 as a straight line with slope $1/k_{\text{BRF}}$, while resistance to LR diffusion across the LR stagnant film inside the LR micelle plots as a straight line with slope $1/k_{\text{LRF}}$. Note the vertical line at $r_{\text{Transition}}$ in Figure 4.5: the length of $1/k_{\text{Overall}}$ below the horizontal line representing resistance to LR chemical reaction equals the length of $1/k_{\text{Overall}}$ from that same horizontal line to the topmost straight line. In other words, resistance to LR chemical reaction equals resistance to diffusion across both stagnant film layers. To the right of $r_{\text{Transition}}$, the semi-batch process is film diffusion rate limited; to the left of $r_{\text{Transition}}$, the semi-batch process is reaction rate limited. We want to operate the semi-batch process to the left of $r_{\text{Transition}}$. Therefore, we will operate the semi-batch process at the highest possible agitation rate and the smallest possible micelle radius.

We obtain the smallest micelle radius by pumping LR through a sintered metal frit. We keep this sintered frit above the BR liquid level and spray LR into the process headspace if the reaction forms a highly

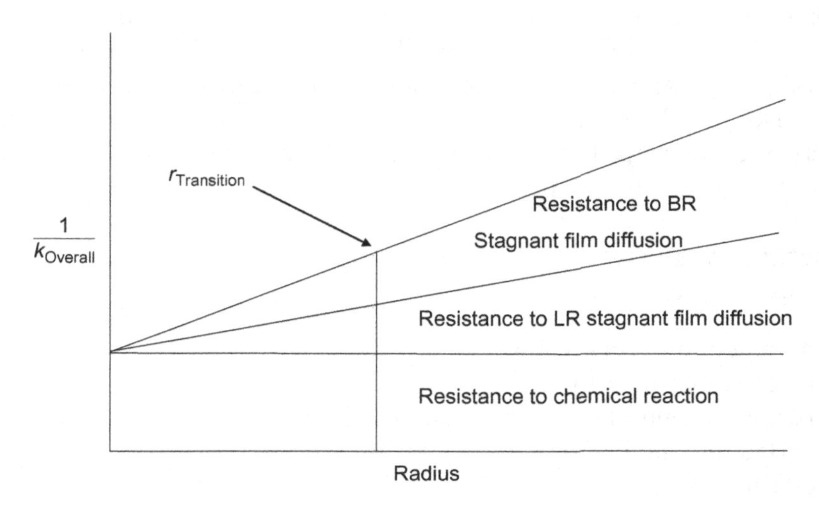

Figure 4.5 Schematic representation of resistance to reactant consumption in semi-batch processes.

viscous, a waxy, or a polymeric product; otherwise, the sintered frit will become plugged with product. Cleaning a plugged, sintered frit is nearly impossible; it is best simply to replace it. If the product has low viscosity and flows freely, then we put the sintered frit in the BR liquid, generally near or under the propeller or turbine agitator. However, when we discharge product from the reactor, we must purge an inert gas through the sintered frit to ensure that it is clear of liquid, otherwise it may become fouled in some manner.

We maintain the smallest micelle radius by guarding against micelle coalescence. Micelle coalescence increases the radius of a micelle, which slows the conversion of reactant to product. We can maintain a small micelle radius by operating the semi-batch process at a low micelle distribution, i.e., operate the process "starved" for LR. This operational mode has the disadvantage of reducing the production capacity of the unit. Using a surfactant may be an option for maintaining a high micelle distribution in the semi-batch process. A surfactant may also improve transfer of reactant from one phase to the other phase.

The effective agitation rate depends upon the fluid volume of the semi-batch process. At constant density, that volume increases as

$$V_R = V_0 + Q_{\mathrm{Feed}}t$$

where V_R is the semi-batch process volume; V_0 is the starting liquid volume in the semi-batch process; Q_{Feed} is the volumetric flow rate of LR into the process; and t is the time duration of LR flow into the process. We install multiple propellers or turbines along the agitator shaft to maintain quality agitation in the process as its volume increases. As the liquid surface rises in the process, it sequentially submerges each propeller or turbine, thereby maintaining the agitation quality required by the process.

Thermal Analysis

In this case, we assume

- no reaction occurs in LR;
- reaction occurs only in BR;
- product accumulates in BR and not in LR;
- perfect mixing in BR;
- constant density;
- constant pressure.

We write the thermal balance for the miscible BR and P phase since it is the continuous phase contacting the heat transfer surfaces; that thermal balance is

$$(n_{BR}c_{P/BR} + n_{P}c_{P/P})\frac{dT}{dt}$$

$$= \Delta H_{BR}R_{BR}(V_0 + Q_{Feed}t) + h_{Interface}a_{BRF}(V_0 + Q_{Feed}t)(T - T_{Feed})$$

$$- U\left(\frac{S_{HeatTransfer}}{V_{Liquid}}\right)(V_0 + Q_{Feed}t)(T - T_{Jacket})$$

where n_{BR} and n_P are the respective moles of BR and moles of P; $c_{P/BR}$ and $c_{P/P}$ are the respective heat capacities of BR and P ($m^2kg/mol * K * s^2$); T is the process temperature (K); t is the time (s); $(\Delta H_{BR})_T$ is the heat of reaction for BR at temperature T ($m^2kg/mol * s^2$); R_{BR} is the kinetic reaction rate of BR ($mol/m^{3}*s$); V_0 is the initial volume of BR in the reactor (m^3); Q_{Feed} is the LR volumetric flow rate (m^3/s); $h_{Interface}$ is the interfacial heat transfer coefficient ($kg/K*s^3$); a_{BRF} is the surface area to volume of the stagnant film surrounding each LR micelle (m^2/m^3); T_{Feed} is the feed temperature of LR (K); U is the overall heat transfer coefficient ($kg/K*s^3$); $S_{HeatTransfer}/V_{Liquid}$ is the heat transfer surface area to liquid volume, i.e., reaction volume (m^2/m^3); and T_{Jacket} is the temperature (K) of the heat transfer fluid flowing through the reactor jacket. The unit for each of the above terms is $m^{2}*kg/s^3$, which is the unit for power.

The thermal balance for LR is

$$(n_{LR}c_{P/LR})\frac{dT}{dt} = Q_{Feed}[LR]c_{P/LR}(T_{Feed} - T)$$
$$- h_{Interface}a_{LRF}(V_0 + Q_{Feed}t)(T - T_{Feed})$$

where n_{LR} is the moles of LR; $c_{P/LR}$ is the LR heat capacity ($m^2kg/mol * K * s^2$); T is the process temperature (K); t is the time (s); Q_{Feed} is the LR volumetric flow rate (m^3/s); [LR] feed concentration (mol/m^3); T_{Feed} is the feed temperature of LR (K); $h_{Interface}$ is the interfacial heat transfer coefficient ($kg/K*s^3$); a_{LRF} is the surface area to volume of the stagnant film inside each LR micelle (m^2/m^3).[1(p400)]

The unit for each of the above terms is $m^{2}*kg/s^3$, which is the unit for power. When integrated, the unit for each term becomes $m^{2}*kg/s^2$, i.e., the unit of energy or work. Note that we neglected the energy added to the process by agitation, and we neglected the work done by the process as its fluid volume expands at constant pressure against the inert gas in the process headspace.

The above component mass and energy balances comprise the design equations for an immiscible liquid–liquid, semi-batch process. Today, we solve them numerically; yesteryear, we simplified and solved them analytically—if we could do so. While we may desire a complete numerical solution, many times we do not have the resources or the time to achieve such a rigorous solution. When we cannot expend the effort to obtain a complete numerical solution, we resort to our yesteryear methods: we make all the pertinent assumptions we can to simplify the component mass and energy balances, then solve them analytically.[1(p400)]

GAS–LIQUID SEMI-BATCH PROCESSES

Kinetic Analysis

When we react a gas with a liquid or react two immiscible liquids, mass transfer occurs simultaneously with reaction. Thus, chemical kinetics is coupled with mass transport, which means we are no longer conducting a reaction, rather, we are managing a complex chemical–physical process. Such processes are usually diffusion rate limited, especially if the gas is sparingly soluble in the liquid or if one liquid is sparingly soluble in a second liquid.

Figure 4.6 presents the macroscopic features of a gas–liquid or immiscible liquids semi-batch process. We completely charge the liquid reactant, either pure or dissolved in an inert solvent, to a reactor equipped with a gas sparger located near or under the agitator. The agitator can be a propeller or one of the many turbine designs available to us. We generally install four wall baffles at 90° separation in the reactor to suppress vortex formation in the liquid during agitation. We use the agitation and baffles to evenly distribute the gas reactant throughout the liquid reactant.

Figure 4.7 shows the microscopic features of a gas–liquid process. Each gas bubble contains gaseous reactant. If the bulk fluid vaporizes into the gas bubbles or if the gas is diluted with an inert gas, then a stagnant film of gas molecules exists adjacent to the gas–liquid

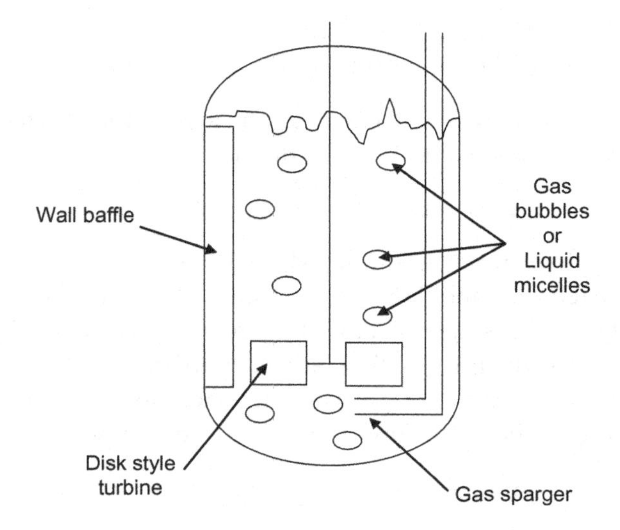

Figure 4.6 Macroscopic features of a gas–liquid semi-batch process.

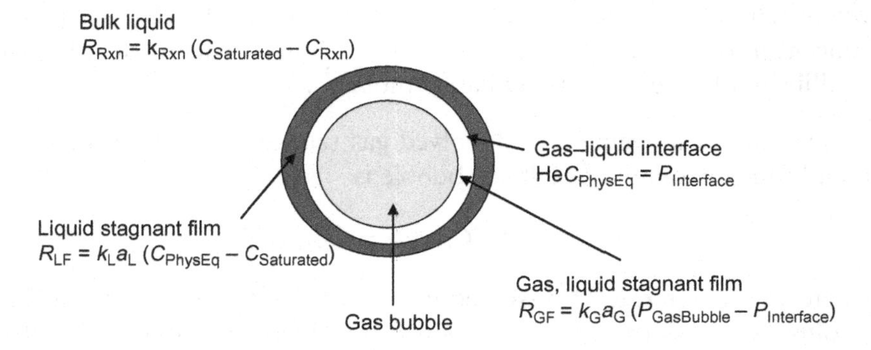

Figure 4.7 Microscopic features of gas–liquid or immiscible liquids semi-batch process.

interface even though some circular motion of gas molecules occurs inside the gas bubble.[13] A stagnant film inside the gas bubble does not form if the bulk fluid does not volatilize into the gas bubbles or if the gas is pure. For this analysis, we assume a stagnant film does form inside the gas bubble.[4] Diffusion governs the movement of gas molecules within this stagnant gas film. Physical equilibrium, expressed as Henry's Law, controls the movement of gas reactant from the gas phase to the liquid phase. A stagnant liquid film surrounds each gas–liquid interface and diffusion governs the movement of solubilized gas within it. The bulk phase of the liquid reactant comprises the fluid moving between each gas bubble. The agitator provides the energy necessary to induce liquid motion within the reactor.

The rate of movement of gas reactant across the stagnant gas film within the gas bubble is mathematically described as

$$R_{GF} = k_G a_G (P_{GasBubble} - P_{Interface})$$

where R_{GF} is the rate of gas molecules moving across the stagnant gas film inside each gas bubble (kg/m*s^3). k_G is the velocity at which gas molecules cross the stagnant gas film (m/s); a_G is the surface area to volume of the stagnant gas film adjacent to the gas–liquid interface (m^2/m^3); $P_{GasBubble}$ and $P_{Interface}$ are the pressures of the gas bubble and the gas reactant at the gas–liquid interface, respectively (kg/m*s^2).

Henry's Law controls the amount of gas dissolved by the liquid reactant; it is

$$\mathrm{He}\, C_{PhysEq} = P_{Interface}$$

where He is the Henry's constant (kg*m^2/mol*s^2); C_{PhysEq} is the concentration of dissolved gas reactant in liquid reactant at physical equilibrium between the two phases (mol/m^3).

The rate of movement of dissolved gas reactant across the stagnant liquid film surrounding each gas bubble is

$$R_{LF} = k_L a_L (C_{PhysEq} - C_{Saturated})$$

where R_{LF} is the rate of dissolved gas moving across the stagnant film (mol/m^3*s); k_L is the velocity at which dissolved gas moves across the stagnant liquid film (m/s); a_L is the surface area to volume of the stagnant liquid film adjacent to the gas–liquid interface (m^2/m^3); and $C_{Saturated}$ is the concentration of dissolved gas in bulk liquid (mol/m^3).

The rate of chemical reaction within the bulk liquid is

$$R_{Rxn} = k_{Rxn}(C_{Saturated} - C_{Rxn})$$

where R_{Rxn} is the rate of chemical reaction (mol/m^3*s); k_{Rxn} is the kinetic rate constant (s^{-1}); and C_{Rxn} is the concentration of dissolved gas reactant at process conditions (mol/m^3).

Unfortunately, in the above rate equations, we only know $P_{GasBubble}$ and C_{Rxn}. $P_{GasBubble}$ is the pressure of the gas as it enters the semi-batch reactor and C_{Rxn} is the dissolved gas in the liquid at process conditions, which we measure by removing a sample from the reactor. Thus, we must reduce the above set of rate equations to one equation containing $P_{\mathrm{GasBubble}}$ and C_{Rxn}. Substituting Henry's Law for $P_{\mathrm{Interface}}$ gives us

$$R_{\mathrm{GF}} = k_{\mathrm{G}}a_{\mathrm{G}}(P_{\mathrm{GasBubble}} - \mathrm{He}C_{\mathrm{PhysEq}})$$

or

$$\frac{R_{\mathrm{GF}}}{\mathrm{He}k_{\mathrm{G}}a_{\mathrm{G}}} = \left(\frac{P_{\mathrm{GasBubble}}}{\mathrm{He}} - C_{\mathrm{PhysEq}}\right)$$

Solving the R_{LF} rate equation for C_{PhysEq}, then substituting into the above equation provides

$$\frac{R_{\mathrm{GF}}}{\mathrm{He}k_{\mathrm{G}}a_{\mathrm{G}}} = \left(\frac{P_{\mathrm{GasBubble}}}{\mathrm{He}} - \frac{R_{\mathrm{LF}}}{k_{\mathrm{L}}a_{\mathrm{L}}} - C_{\mathrm{Saturated}}\right)$$

Rearranging the above equation gives

$$\frac{R_{\mathrm{GF}}}{\mathrm{He}k_{\mathrm{G}}a_{\mathrm{G}}} + \frac{R_{\mathrm{LF}}}{k_{\mathrm{L}}a_{\mathrm{L}}} = \left(\frac{P_{\mathrm{GasBubble}}}{\mathrm{He}} - C_{\mathrm{Saturated}}\right)$$

Solving the R_{Rxn} rate equation for $C_{\mathrm{Saturated}}$, then substituting into the last equation above yields

$$\frac{R_{\mathrm{GF}}}{\mathrm{He}k_{\mathrm{G}}a_{\mathrm{G}}} + \frac{R_{\mathrm{LF}}}{k_{\mathrm{L}}a_{\mathrm{L}}} = \left(\frac{P_{\mathrm{GasBubble}}}{\mathrm{He}} - \frac{R_{\mathrm{Rxn}}}{k_{\mathrm{Rxn}}} - C_{\mathrm{Rxn}}\right)$$

and upon rearranging gives

$$\frac{R_{\mathrm{GF}}}{\mathrm{He}k_{\mathrm{G}}a_{\mathrm{G}}} + \frac{R_{\mathrm{LF}}}{k_{\mathrm{L}}a_{\mathrm{L}}} + \frac{R_{\mathrm{Rxn}}}{k_{\mathrm{Rxn}}} = \left(\frac{P_{\mathrm{GasBubble}}}{\mathrm{He}} - C_{\mathrm{Rxn}}\right)$$

We now have all the rates related to the two variables we can measure. We know the overall rate of the process is

$$R_{\mathrm{Overall}} = k_{\mathrm{Overall}}\left(\frac{P_{\mathrm{GasBubble}}}{\mathrm{He}} - C_{\mathrm{Rxn}}\right)$$

where R_{Overall} is the overall rate (mol/m^3*s) for the semi-batch process and k_{Overall} is the overall rate constant for the semi-batch process (s^{-1}).

Since none of the rates can be faster than the slowest rate and the overall rate equals the slowest rate, we can write

$$\frac{R_G}{\mathrm{He}} = R_{LF} = R_{Rxn} = R_{Overall}$$

Therefore

$$\frac{R_{Overall}}{k_G a_G} + \frac{R_{Overall}}{k_L a_L} + \frac{R_{Overall}}{k_{Rxn}} = \left(\frac{P_{GasBubble}}{\mathrm{He}} - C_{Rxn}\right)$$

or

$$R_{Overall}\left(\frac{1}{k_G a_G} + \frac{1}{k_L a_L} + \frac{1}{k_{Rxn}}\right) = \left(\frac{P_{GasBubble}}{\mathrm{He}} - C_{Rxn}\right)$$

Rearranging the above equation gives us

$$R_{Overall} = \frac{1}{((1/k_G a_G) + (1/k_L a_L) + (1/k_{Rxn}))}\left(\frac{P_{GasBubble}}{\mathrm{He}} - C_{Rxn}\right)$$

which means

$$k_{Overall} = \frac{1}{((1/k_G a_G) + (1/k_L a_L) + (1/k_{Rxn}))}$$

or

$$\frac{1}{k_{Overall}} = \frac{1}{k_G a_G} + \frac{1}{k_L a_L} + \frac{1}{k_{Rxn}}$$

The above equation represents serial resistance in a semi-batch process. If

$$k_G a_G \text{ and } k_L a_L \gg k_{Rxn}$$

then the semi-batch process is reaction rate limited. If

$$k_{Rxn} \gg k_G a_G \text{ or } k_L a_L$$

then the semi-batch process is diffusion rate limited.

For a semi-batch process to be diffusion rate limited, either

$$k_{Rxn} \gg k_G a_G$$

or

$$k_{Rxn} \gg k_L a_L$$

holds true. The question is: which stagnant film is limiting the rate of the semi-batch process? For spheres, the Sherwood numbers for gas molecules and liquid molecules are of similar order of magnitude. Thus

$$O\left(\frac{k_G \delta_G}{D_G}\right) = O\left(\frac{k_L \delta_L}{D_L}\right)$$

where k_G is the rate constant for molecules crossing a stagnant film (m/s); δ is the thickness of the stagnant film (m); D is the diffusivity of the respective phase (m^2/s); and $O(\ )$ signifies order of magnitude. We know that $D_G \gg D_L$; therefore, $k_G \delta_G \gg k_L \delta_L$. If $\delta_G = \delta_L$, then $k_G \gg k_L$. Therefore, the movement of dissolved gas across the stagnant liquid film is most likely the rate limiting step in the semi-batch process.

The serial resistance for the semi-batch process is

$$\frac{1}{k_{\text{Overall}}} = \frac{1}{k_G a_G} + \frac{1}{k_L a_L} + \frac{1}{k_{\text{Rxn}}}$$

If we consider k_G, a_G, and k_{Rxn} to be constant relative to $k_L a_L$, then we can combine them into one term. Multiplying the first term left of the equal sign by 1; that is, by $k_{\text{Rxn}}/k_{\text{Rxn}}$, then multiply the left-most term by 1; or, $k_G a_G / k_G a_G$ yields

$$\frac{1}{k_{\text{Overall}}} = \frac{k_{\text{Rxn}}}{k_G a_G k_{\text{Rxn}}} + \frac{1}{k_L a_L} + \frac{k_G a_G}{k_{\text{Rxn}} k_G a_G}$$

Combining terms gives us

$$\frac{1}{k_{\text{Overall}}} = \frac{k_{\text{Rxn}} + k_G a_G}{k_G a_G k_{\text{Rxn}}} + \frac{1}{k_L a_L}$$

which we recognize as the equation for a straight line if we define the ordinate as $1/k_{\text{Overall}}$ and the abscissa as $1/k_L$ or as $1/a_L$. The intercept for this straight line is $(k_{\text{Rxn}} + k_G a_G)/k_G a_G k_{\text{Rxn}}$.

A relationship between k_{Overall} and $k_L a_L$ has been intuitively surmised since the formation of chemical engineering as a separate engineering discipline. One semi-batch variable available to the early chemical engineer was agitator speed or agitator RPMs. By increasing agitator RPM, pioneer chemical engineers found that k_{Overall} increased, then plateaued. Figure 4.8 shows this common result. We generally say a semi-batch process demonstrating a linear response between k_{Overall}

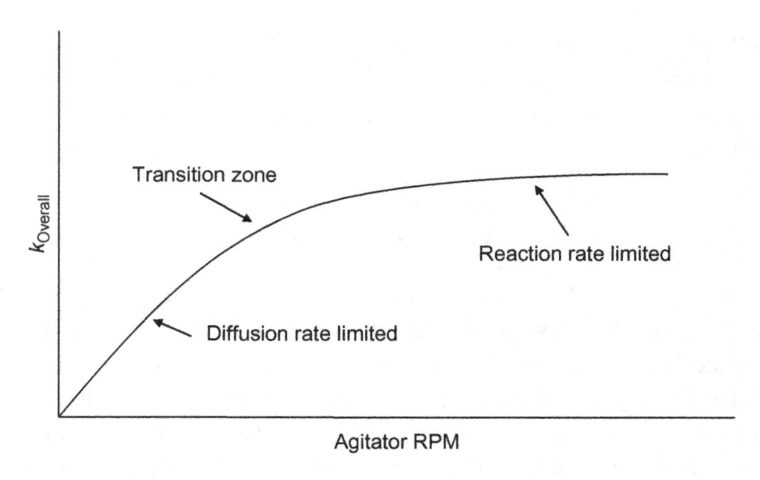

Figure 4.8 $k_{Overall}$ *as a function of agitator RPMs.*

and agitator RPM is "diffusion" rate limited; we say a plateaued response represents a "reaction" rate limited semi-batch process.

Unfortunately, agitator RPM contains little information, except for the semi-batch reactor from which it was measured. The reason: no two semi-batch reactors have the same geometry; thus, they violate geometric similarity.[14–16] Batch and semi-batch reactors differ with respect to the

- heat transfer coils, both number and configuration;
- thermowells and other piping;
- baffle configurations.

extending into the agitated liquid, all of which sums to no two batch or semi-batch reactors being identical. Therefore, at a given RPM, the motion of the liquid within two batch or semi-batch reactors is different. And, the method of agitation is different from one semi-batch reactor to another. The agitator can be a propeller, a flat-blade turbine, or a pitched-blade turbine, all of which impart a different motion to the liquid in a batch or semi-batch reactor. In other words, the semi-batch process in each reactor is different.

To overcome this problem, Rushton et al. proposed using power transferred to the liquid in a semi-batch reactor as a scalable parameter.[17,18] They called this parameter the "power number," which is expressed in general terms as

$$\mathrm{PN} = \frac{Pg_c}{\rho N^3 D^5} = \kappa \left(\frac{\rho N D^2}{\mu}\right)^a \left(\frac{N^2 D}{g}\right)^b \left(\frac{T}{D}\right)^c \left(\frac{C}{D}\right)^d \left(\frac{H}{D}\right)^e \left(\frac{p}{D}\right)^f$$
$$\left(\frac{W}{D}\right)^g \left(\frac{L}{D}\right)^h \left(\frac{n}{n_{\mathrm{Ref}}}\right)^i \left(\frac{n_{\mathrm{Baf}}}{n_{\mathrm{BafRef}}}\right)^j \cdots$$

where PN is the power number; P, power; g_c, the gravitational constant; ρ, liquid density; N, agitator rotation speed; D, agitator diameter; κ, a constant; μ, liquid viscosity; g, gravitational acceleration; T, reactor diameter; C, clearance of agitator from reactor bottom; H, liquid height; p, agitator blade pitch; W, agitator blade width; L, agitator blade length; n, number of agitator blades; n_{Ref}, number of agitator blades of reference geometry; n_{Baf}, number of baffles; n_{BafRef}, number of baffles of reference geometry. The exponents a, b, c, d, and so on are determined experimentally.

The above dimensionless equation requires comment. First, the number of variables in it depends upon the inventiveness of the chemical engineer. Second, power transferred to the liquid in a semi-batch reactor is not constant. It changes as the agitator bearing ages. Third, its basic form may have been derived using the methods of dimensional analysis current in 1950, but most of the ratios have simply been added to the original equation. Fourth, the number of terms included or deleted from this equation depends upon the design engineer. In other words, the above equation can be rather arbitrary and misleading.[14(chp 2)]

Despite the issues surrounding the power number, as written above, chemical engineers still use it as a semi-batch process parameter. Figure 4.8 shows that k_{Overall} depends upon agitator RPM. Thus, k_{Overall} depends in some manner on the stagnant liquid film surrounding each bubble since increasing RPM presumably increases the velocity of the bulk fluid flowing between bubbles. It is well established by fluid dynamics that the thickness of a stagnant gas or liquid film adjacent to a surface decreases as the velocity of the fluid flowing over it increases. Mathematically

$$\delta \propto \frac{1}{\sqrt{Re}} = \frac{1}{\sqrt{\rho D v / \mu}} = \sqrt{\mu / \rho D v}$$

where Re is the Reynolds number in the direction of fluid flow; ρ is the density of the flowing fluid; D is the diameter of the conduit through which the fluid flows; v is the velocity of the flowing fluid; and μ is the viscosity of the flowing fluid.[19] If we postulate each gas bubble to be spherical, then a_L is

$$a_L = \frac{4\pi\delta^2}{(4/3)\pi\delta^3} = \frac{3}{\delta}$$

Thus, $k_{Overall}$ is a function of $k_L a_L$ and that $k_L a_L$ is a function of power number. In other words

$$k_{Overall} = f(k_L a_L)$$
$$k_L a_L = f(PN)$$

But, $k_L a_L$ is rarely, if ever, presented solely as a function of power number. Chemical engineers prefer to describe the quality of agitation as power input per volume. In many cases, they express $k_L a_L$ as a function of power to volume ratio, liquid fluid properties, and gas bubble coalescence properties;[14(p496)] for example as

$$k_L a_L = f(P_T/V, \text{liquid fluid properties, gas bubble coalescence properties})$$

where P_T is the power transmitted to the liquid fluid in a semi-batch reactor or tank and V is the liquid volume in the semi-batch reactor. The majority of such functions have simpler forms since gas coalescence property data is not common. Many $k_L a_L$ functions are written in terms of the power contributed to the liquid fluid by the injected gas; for example

$$k_L a_L = \kappa \left(\frac{P_G}{V}\right)^a v_S^b$$

where P_G is the power contributed to the liquid fluid by the injected gas; V, the liquid volume in the semi-batch reactor; and v_S, the superficial gas velocity. κ, a, and b are experimentally determined constants and κ has dimensions necessary to balance the equation.[14(p483 and 491)]

Proposing functions and developing empirical relations for $k_L a_L$ has been a burgeoning cottage industry since the early 1950s. Many articles are published annually upon this subject, as well as chapters in agitation and mixing tomes. Such published functions and empirical correlations may be valid for the semi-batch reactor from which they are derived, but, because no two semi-batch reactors are geometrically similar, these correlations are not valid, except in a grossly general manner, for any other semi-batch reactor.

With the above information, let us reconsider the resistance equation for a semi-batch process; it is

$$\frac{1}{k_{\mathrm{Overall}}} = \frac{k_{\mathrm{Rxn}} + k_{\mathrm{G}}a_{\mathrm{G}}}{k_{\mathrm{G}}a_{\mathrm{G}}k_{\mathrm{Rxn}}} + \frac{1}{k_{\mathrm{L}}a_{\mathrm{L}}}$$

If we assume it describes a straight line, then the intercept is $k_{\mathrm{Rxn}} + k_{\mathrm{G}}a_{\mathrm{G}}/k_{\mathrm{G}}a_{\mathrm{G}}k_{\mathrm{Rxn}}$ and its slope is either $1/k_{\mathrm{L}}$ or $1/a_{\mathrm{L}}$. If we postulate each gas bubble to be spherical, then a_{L} is

$$a_{\mathrm{L}} = \frac{4\pi r^2}{(4/3)\pi r^3} = \frac{3}{r}$$

Thus, a_{L} increases as r decreases, which it does over time as gas dissolves into the liquid. In other words, a_{L} is not constant—it changes as time progresses. a_{L} cannot be the slope of this straight line; rather, it is the independent variable in this relationship.

Now consider k_{L}: it depends upon the mean free path of the dissolved gas in the liquid and the mean free path depends upon liquid density. We can, therefore, consider k_{L} to be constant within the variability of liquid density.[14(p483)] Thus, $1/k_{\mathrm{L}}$ is the slope of the resistance equation describing a semi-batch process. In summary, all the proposed functions relating $k_{\mathrm{L}}a_{\mathrm{L}}$ to a variety of independent variables are actually functions relating a_{L} to those same independent variables. This knowledge allows us to refocus our effort to improve a semi-batch process, i.e., to improve k_{Overall}.

If our goal is to improve k_{Overall} for a semi-batch process, then we need to increase a_{L}, which means producing the smallest possible bubble in the liquid phase of the process. Producing small bubble sizes within a process liquid via agitation is expensive and the surface tension of the process liquid and the maximum attainable RPMs produced place a limit on bubble size. The most cost effective and safest location for producing the smallest possible bubble, thus the largest a_{L}, is at the point where we inject gas into the liquid. Thus, the inlet gas nozzle for a semi-batch process should be a metal frit which produces the appropriate-sized bubble and a_{L}. Determining bubble size and a_{L} at the inlet gas nozzle allows us to redefine the purpose of agitation. The primary goal of agitation is now one of maintaining an even distribution of bubbles throughout the process liquid. The secondary goal of agitation is to minimize bubble coalescence: we do not want small bubbles coalescencing with other bubbles to form larger bubbles, thereby reducing a_L and consequently lowering $k_{Overall}$. Redefining our agitation requirement in this manner reduces operating costs, i.e., the agitator does not consume as much energy as when we operate it to create bubbles or reduce bubble size, and it reduces maintenance costs, i.e., agitator bearings do not wear as quickly as when we operate the agitator at high RPMs.

Using a metal frit in a gas–liquid semi-batch process has its disadvantages. First, after discharging the reactor, gas must be blown through the metal frit to remove any residual liquid from it, especially if reaction can occur in the residual liquid. Such reaction can foul and plug the metal frit. Second, a wobbly agitator can shear the metal frit off the inlet piping, thereby causing untold problems within the reactor. Third, a tool dropped through a manway when the reactor is empty can strike the metal frit and damage it or shear it off the inlet piping. And, lastly, while retrieving that dropped tool or repairing the interior of the reactor, an operator can step on the metal frit and damage it or shear it off the inlet piping. We can overcome these disadvantages with diligent training. The payoff for such diligence is rapid and significant.

Thermal Analysis

In this case, we assume

- no reaction occurs in the gas phase;
- reaction occurs only in the liquid phase;
- product accumulates in the liquid phase;
- perfect mixing of the liquid phase;
- constant density;
- constant volume;
- constant pressure.

The thermal balance for the liquid phase is

$$(n_{Liq}c_{P/Liq} + n_P c_{P/P})\frac{dT}{dt} = (\Delta H_{Liq})_T R_{Liq} + h_{Interface}a_{Liq}(T - T_{Gas}) - UA_{HeatTransfer}(T - T_{Jacket})$$

where n_{Liq} is the moles of liquid; $c_{P/Liq}$ is the liquid heat capacity (m^2*kg/mol*K*s^2); T is the process temperature (K); t is the time (s); $(\Delta H_{Liq})_T$ is the heat of reaction for liquid at temperature T (m^2*kg/mol*s^2); R_{Liq} is the kinetic reaction rate of liquid (mol/m^3*s); T_{Gas} is the gas temperature (K); U is the overall heat transfer coefficient (kg/K*s^3); $A_{HeatTransfer}$ is the heat transfer surface area (m^2); and T_{Jacket} is the temperature (K) of the heat transfer fluid flowing through the reactor jacket. The units for each of the above terms are m^{2*}kg/s^3, which are the units for power.

The thermal balance for the gas is

$$(n_{Gas}c_{P/Gas})\frac{dT}{dt} = Q_{Feed}C_{Gas}c_{P/Gas}(T_{Gas} - T) - h_{Interface}a_{GasFilm}(T - T_{Feed})$$

where n_{Gas} is the moles of gas; $c_{P/Gas}$ is the gas heat capacity (m^{2*}kg/mol*K*s^2); T is the process temperature (K); t is the time (s); Q_{Feed} is the gas volumetric flow rate (m^3/s); C_{Gas} is the gas feed concentration (mol/m^3); T_{Gas} is the feed temperature of the gas (K); $h_{Interface}$ is the interfacial heat transfer coefficient (kg/K*s^3); $a_{GasFilm}$ is the surface area to volume of the stagnant film inside each gas bubble (m^2/m^3).[1(p400)] The unit for each of the above terms is m^{2*}kg/s^3, which is the unit for power. When integrated, the unit for each term becomes m^{2*}kg/s^2, i.e., the unit of energy or work. Note that we neglected the energy added to the process by agitation, and we neglected the work done by the process as its fluid volume expands at constant pressure against the inert gas in the process headspace.

These energy balances, along with the appropriate component balances for the process, comprise the design equations for a gas–liquid, semi-batch process.[1(p400)]

SUMMARY

This chapter discussed the reasons for choosing a semi-batch reactor rather than a batch reactor. It then presented the design equations for conducting a miscible liquid–liquid reaction in a semi-batch reactor. This chapter then reviewed the design equations for an immiscible liquid–liquid process in a semi-batch reactor, and it concluded with a discussion of the design equations for a gas–liquid process conducted in a semi-batch reactor.

REFERENCES

1. Nauman E. *Chemical reactor design, optimization, and scaleup*. 2nd ed. New York, NY: John Wiley & Sons; 2008. p. 71–72.

2. Smith J. *Chemical engineering kinetics*. 3rd ed. New York, NY: McGraw-Hill Book Company; 1983. p. 83.
3. Carberry J. *Chemical and catalytic reaction engineering*. New York, NY: McGraw-Hill Book Company; 1976. p. 26.
4. Harriott P. *Chemical reactor design*. New York, NY: Marcel Dekker, Inc.; 2003. p. 5.
5. Rase H. *Chemical reactor design for process plants: volume 1—principles and techniques*. New York, NY: John Wiley & Sons; 1977.
6. Davis M, Davis R. *Fundamentals of chemical reactor engineering*. New York, NY: McGraw-Hill Book Company; 2003. p. 93.
7. Froment G, Bischoff K. *Chemical reactor analysis and design*. New York, NY: John Wiley & Sons; 1979. p. 411.
8. Worstell J. Don't act like a novice about reaction engineering. *Chem Eng Prog* 2001;**97** (3):68–72.
9. Coker A. *Modeling of chemical kinetics and reactor design*. Boston, MA: Gulf Professional Publishing; 2001. p. 306–312.
10. Butt J. *Reaction kinetics and reactor design*. 2nd ed. Marcel Dekker, Inc.; 2000. p. 274–277.
11. Miller J. *Separation methods in chemical analysis*. New York, NY: John Wiley & Sons; 1975. p. 6.
12. Treybal R. *Mass transfer operations*. New York, NY: McGraw-Hill Book Company; 1955. p. 383–391.
13. Clift R, Grace J, Weber M. *Bubbles, drops, and particles*. Mineola, NY: Dover Publications, Inc.; 2005chapter 1 (originally published by Academic Press, Inc., New York, NY, 1978).
14. Tatterson G. *Fluid mixing and gas dispersion in agitated tanks*. Second Printing. Tatterson G, editor. Greensboro, NC: 1991 [chapters 1–2].
15. Tatterson G. *Scaleup and design of industrial mixing processes*. New York, NY: McGraw-Hill Book Company; 1994 [chapters 1–2].
16. Worstell J. *Dimensional analysis*. Waltham, MA: Butterworth-Heinemann; 2014 [chapter 8].
17. Rushton J. Applications of fluid mechanics and similitude to scaleup problems: part 1. *Chem Eng Prog* 1952;**48**(1):33–8.
18. Rushton J. Applications of fluid mechanics and similitude to scaleup problems: part 2. *Chem Eng Prog* 1952;**48**(2):95–102.
19. Granger R. *Fluid mechanics*. New York, NY: Dover Publications, Inc.; 1995. p. 708 (originally published by Holt, Reinhart and Winston, New York, NY, 1985).

CHAPTER 5

Batch and Semi-Batch Operations

PROCESS CYCLE TIME

We must know the process cycle time for a batch or semi-batch process in order to estimate the number of reactors required for a given plant capacity and to estimate the staffing requirement for that plant. By process cycle time, we mean the time between initiating one reaction and the next, subsequent reaction. One process cycle for a batch or semi-batch process involves

- charging the reactor with feed;
- heating the reactor contents to a specified reaction temperature;
- completing the reaction;
- cooling the reactor contents to a specified discharge temperature;
- discharging product from the reactor;
- flushing, i.e., cleaning, the reactor;
- drying the reactor;
- preparing the reactor for the next charge.

We must estimate the time required for each of the above steps before finalizing our plant design.

We can easily calculate the time to charge a batch or semi-batch reactor by initially specifying a reactor size and a feed volumetric flow rate. Therefore, if we specify a 3000 gallon (11.4 m^3) batch reactor and a volumetric flow rate of 100 gallons/min (0.379 m^3/min), then the time required to fill the reactor 80% full with reactant is

$$t = \frac{V_{\text{Liquid}}}{Q_{\text{Feed}}} = \frac{0.8\ (3000\ \text{gallons})}{100\ \text{gallons/min}} = 24\ \text{min}$$

where t is the fill time.

After charging the feed to the batch or semi-batch reactor, we begin heating it to reaction temperature. Consider a batch reactor containing V_{Liquid} with a heat capacity of $c_{\text{P/Liq}}$ at an initial temperature T_{Initial}.

Our batch reactor contains a heat exchanger in the form of coiled loops through which an isothermal condensing fluid flows, thereby heating the liquid contents of the reactor. The heat accumulating in the reactor's contents equals the rate of heat transfer from the coiled loops; that is

$$V_{\text{Liquid}}\rho_{\text{Liq}}c_{\text{P/Liq}}\left(\frac{\mathrm{d}T}{\mathrm{d}t}\right) = UA(T_{\text{HeatFluid}} - T)$$

where V_{Liquid} is the liquid volume in the reactor (m^3); ρ_{Liq} is the density of the liquid in the reactor (kg/m^3); $c_{\text{P/Liq}}$ is the specific heat capacity of the liquid in the reactor ($m^2/K{*}s^2$); T is the temperature of the liquid in the reactor (K); t is the time (s); U is the overall heat transfer coefficient for the coiled loops in the reactor ($kg/K{*}s^3$); A is the heat transfer surface area of the coiled loops (m^2); and $T_{\text{HeatFluid}}$ is the temperature of the heating medium flowing through the coiled loops (K). Rearranging gives

$$\frac{\mathrm{d}T}{\mathrm{d}t} = \left(\frac{UA}{V_{\text{Liquid}}\rho_{\text{Liq}}c_{\text{P/Liq}}}\right)(T_{\text{HeatFluid}} - T)$$

Then, separating variables and integrating yields

$$\int_{T_{\text{Initial}}}^{T_{\text{Reaction}}} \frac{\mathrm{d}T}{(T_{\text{HeatFluid}} - T)} = \left(\frac{UA}{V_{\text{Liquid}}\rho_{\text{Liq}}c_{\text{P/Liq}}}\right)\int_0^t \mathrm{d}t$$

$$-\ln\left[\frac{T_{\text{HeatFluid}} - T_{\text{Reaction}}}{T_{\text{HeatFluid}} - T_{\text{Initial}}}\right] = \left(\frac{UA}{V_{\text{Liquid}}\rho_{\text{Liq}}c_{\text{P/Liq}}}\right)t$$

Finally, moving the negative sign inside the logarithm function gives

$$\ln\left[\frac{T_{\text{HeatFluid}} - T_{\text{Initial}}}{T_{\text{HeatFluid}} - T_{\text{Reaction}}}\right] = \left(\frac{UA}{V_{\text{Liquid}}\rho_{\text{Liq}}c_{\text{P/Liq}}}\right)t$$

Solving for time t yields

$$\left(\frac{V_{\text{Liquid}}\rho_{\text{Liq}}c_{\text{P/Liq}}}{UA}\right)\ln\left[\frac{T_{\text{HeatFluid}} - T_{\text{Initial}}}{T_{\text{HeatFluid}} - T_{\text{Reaction}}}\right] = t$$

This equation gives the time required to heat the liquid volume in the batch reactor from its initial temperature to the specified reaction temperature.[1,2] The derivation is the same for a jacketed batch reactor; therefore, the above equation is valid for a batch reactor with coiled loops or with a jacket.

Other heating options are available. For example, what if we replace the isothermal heating fluid with a non-isothermal heating fluid? For the same equipment as above, the heating fluid enters the coiled loop at temperature T_{Inlet}, but exits the coiled loop at T_{Exit}. The heat lost by the heating fluid while passing through the coiled loop equals the heat transferred to the batch reactor; that is

$$Q_{\text{HeatFluid}} = Q_{\text{HeatTransfer}}$$
$$w_{\text{HeatFluid}} c_{\text{P/HeatFluid}}(T_{\text{Inlet}} - T_{\text{Exit}}) = UA(\text{LMTD})$$

where $w_{\text{HeatFluid}}$ is the weight velocity of heating fluid through the coiled loop (kg/s); $c_{\text{P/HeatFluid}}$ is the specific heat capacity of the heating fluid (m^2/K^*s); U and A have been previously defined; and LMTD is the "log mean temperature difference," defined as

$$\text{LMTD} = \frac{T_{\text{Inlet}} - T_{\text{Exit}}}{\ln[T_{\text{Inlet}} - T/T_{\text{Exit}} - T]}$$

where T is the temperature of the liquid inside the batch reactor. The heat balance equation is, therefore

$$w_{\text{HeatFluid}} c_{\text{P/HeatFluid}}(T_{\text{Inlet}} - T_{\text{Exit}}) = UA\left\{\frac{T_{\text{Inlet}} - T_{\text{Exit}}}{\ln[T_{\text{Inlet}} - T/T_{\text{Exit}} - T]}\right\}$$

Note, however, that the temperature of the liquid inside the batch or semi-batch reactor changes with time. The energy balance for the liquid inside the batch or semi-batch reactor is

$$V_{\text{Liq}} \rho_{\text{Liq}} c_{\text{P/Liq}} \left(\frac{dT}{dt}\right) = w_{\text{HeatFluid}} c_{\text{p/HeatFluid}}(T_{\text{Inlet}} - T_{\text{Exit}})$$

We can solve this differential equation by solving the second equation above for T_{Exit}, then substituting T_{Exit} into the differential equation. Doing so yields

$$T_{\text{Exit}} = T + \frac{T_{\text{Inlet}} - T}{e^{(UA/(wc_{\text{P}})_{\text{HeatFluid}})}}$$

and gives

$$V_{\text{Liq}} \rho_{\text{Liq}} c_{\text{P/Liq}} \left(\frac{dT}{dt}\right) = w_{\text{HeatFluid}} c_{\text{p/HeatFluid}}\left(T_{\text{Inlet}} - \left[T + \frac{T_{\text{Inlet}} - T}{e^{(UA/(wc_{\text{P}})_{\text{HeatFluid}})}}\right]\right)$$

Expanding, simplifying, rearranging, then integrating yields

$$\int_{T_{\text{Initial}}}^{T_{\text{Reaction}}} \frac{\mathrm{d}T}{T_{\text{Inlet}} - T} = \left(\frac{w_{\text{HeatFluid}} c_{\text{p/HeatFluid}}}{V_{\text{Liq}} \rho_{\text{Liq}} c_{\text{P/Liq}}}\right)\left(\frac{e^{(UA/(wc_{\text{P}})_{\text{HeatFluid}})} - 1}{e^{(UA/(wc_{\text{P}})_{\text{HeatFluid}})}}\right) \int_0^t \mathrm{d}t$$

The integration gives

$$\ln\left(\frac{T_{\text{Inlet}} - T_{\text{Initial}}}{T_{\text{Inlet}} - T_{\text{Final}}}\right) = \left(\frac{w_{\text{HeatFluid}} c_{\text{p/HeatFluid}}}{V_{\text{Liq}} \rho_{\text{Liq}} c_{\text{P/Liq}}}\right)\left(\frac{e^{(UA/(wc_{\text{P}})_{\text{HeatFluid}})} - 1}{e^{(UA/(wc_{\text{P}})_{\text{HeatFluid}})}}\right) t$$

where we have moved the negative sign from integrating into the logarithm term. The time to heat the liquid inside the batch reactor is then[1,2]

$$\left(\frac{V_{\text{Liq}} \rho_{\text{Liq}} c_{\text{P/Liq}}}{w_{\text{HeatFluid}} c_{\text{p/HeatFluid}}}\right)\left(\frac{e^{(UA/(wc_{\text{P}})_{\text{HeatFluid}})}}{e^{(UA/(wc_{\text{P}})_{\text{HeatFluid}})} - 1}\right) \ln\left(\frac{T_{\text{Inlet}} - T_{\text{Initial}}}{T_{\text{Inlet}} - T_{\text{Final}}}\right) = t$$

It is not uncommon for a batch or semi-batch reactor to possess neither internal coiled loops nor a jacket for heat transfer. For reactors without internal coiled loops or a jacket, we pump the liquid inside the reactor through an external heat exchanger, generally a shell and tube exchanger. We call these external heat exchangers "pump-around loops." We withdraw the liquid inside the reactor from the bottom of the reactor, pump it through a shell and tube heat exchanger, then return it to the top of the reactor, either spraying it into the reactor's headspace or distributing it in the liquid inside the reactor via a submerged pipe and nozzle. For a pump-around loop using an isothermal heating fluid, the energy balance is

$$V_{\text{Liq}} \rho_{\text{Liq}} c_{\text{P/Liq}} \left(\frac{\mathrm{d}T}{\mathrm{d}t}\right) = w_{\text{Liq}} c_{\text{p/Liq}} (T_{\text{Xger}} - T)$$

where V_{Liquid} is the liquid volume in the reactor (m^3); ρ_{Liq} is the density of the liquid in the reactor (kg/m^3); $c_{\text{P/Liq}}$ is the specific heat capacity of the liquid in the reactor (m^2/K*s^2); T is the temperature of the liquid in the reactor (K); t is the time (s); w_{Liq} is the weight flow rate (kg/s); and T_{Xger} is the temperature of the reactor liquid exiting the heat exchanger (K).[1(p628)] The energy balance around the heat exchanger is

$$w_{\text{Liq}} c_{\text{p/Liq}} (T_{\text{Xger}} - T) = UA\left[\frac{T_{\text{Xger}} - T}{\ln(T_{\text{HeatFluid}} - T / T_{\text{HeatFluid}} - T_{\text{Xger}})}\right]$$

where $T_{\text{HeatFluid}}$ is the temperature of the heating fluid entering and exiting the shell and tube heat exchanger; all other variables are defined above. Solving the above equation for T_{Xger} gives

$$T_{\text{Xger}} = T_{\text{HeatFluid}} - \frac{T_{\text{HeatFluid}} - T}{e^{(UA/w_{\text{Liq}} c_{\text{p/Liq}})}}$$

Substituting T_{Xger} into the above differential equation yields

$$V_{\text{Liq}} \rho_{\text{Liq}} c_{\text{P/Liq}} \left(\frac{\text{d}T}{\text{d}t} \right) = w_{\text{Liq}} c_{\text{p/Liq}} \left(T_{\text{HeatFluid}} - \frac{T_{\text{HeatFluid}} - T}{e^{(UA/w_{\text{Liq}} c_{\text{p/Liq}})}} - T \right)$$

which gives, upon rearranging and integrating

$$\ln\left(\frac{T_{\text{HeatFluid}} - T_{\text{Initial}}}{T_{\text{HeatFluid}} - T_{\text{Final}}} \right) = \left(\frac{w_{\text{Liq}}}{V_{\text{Liq}} \rho_{\text{Liq}}} \right) \left(\frac{e^{(UA/w_{\text{Liq}} c_{\text{p/Liq}})} - 1}{e^{(UA/w_{\text{Liq}} c_{\text{p/Liq}})}} \right) t$$

The time to heat the liquid inside the batch reactor is then

$$\left(\frac{V_{\text{Liq}} \rho_{\text{Liq}}}{w_{\text{Liq}}} \right) \left(\frac{e^{(UA/w_{\text{Liq}} c_{\text{p/Liq}})}}{e^{(UA/w_{\text{Liq}} c_{\text{p/Liq}})} - 1} \right) \ln\left(\frac{T_{\text{HeatFluid}} - T_{\text{Initial}}}{T_{\text{HeatFluid}} - T_{\text{Final}}} \right) = t$$

For analysis of more complex heating systems, see the open literature, which is rife with them.[1(p628, chp 18)]

We commence the chemical reaction once the liquid inside a batch or semi-batch reactor reaches reaction temperature. The time required to complete the chemical reaction depends upon the chemical kinetics of the reaction mechanism and the desired final product concentration or reactant concentration. For a first-order reaction, the time required to reach a given product concentration is

$$t = \left(\frac{1}{k} \right) \ln\left[\frac{P_{\text{Final}}}{P_{\text{Initial}}} \right]$$

or to reduce a given reactant to a specified concentration is

$$t = -\left(\frac{1}{k} \right) \ln\left(\frac{[R]_{\text{Final}}}{[R]_{\text{Initial}}} \right) = \left(\frac{1}{k} \right) \ln\left(\frac{[R]_{\text{Initial}}}{[R]_{\text{Final}}} \right)$$

For a second-order chemical reaction, the time to reach a specified final reactant concentration is

$$t = \frac{1}{k} \left(\frac{R_{\text{Initial}} - R_{\text{Final}}}{R_{\text{Final}} R_{\text{initial}}} \right)$$

If the kinetics for the chemical reaction are too complex to be reduced to a manageable equation, then we can plot product concentration or reactant concentration as a function of time and read the time required to reach a specified concentration from the plot (see Chapter 2).

After completing the chemical reaction, we may need to cool the liquid contents of the batch or semi-batch reactor before discharging it. For a batch or semi-batch reactor equipped with internal coiled loops or a jacket containing finished product at temperature $T_{Reaction}$, the time to cool the liquid volume using an isothermal cooling fluid at $T_{CoolFluid}$ is[1]

$$t = \left(\frac{V_{Product}\rho_{Product}c_{P/Product}}{UA}\right)\ln\left[\frac{T_{Reaction} - T_{CoolFluid}}{T_{Final} - T_{CoolFluid}}\right]$$

where all the other variables have been previously defined. For a similar process using a non-isothermal cooling fluid, the time to cool the liquid volume inside the reactor is

$$t = \left(\frac{e^{(UA/w_{CoolFluid}c_{P/CoolFluid})}}{e^{(UA/w_{CoolFluid}c_{P/CoolFluid})} - 1}\right)\left(\frac{V_{Product}\rho_{Product}}{w_{CoolFluid}c_{P/CoolFluid}}\right)\ln\left[\frac{T_{CoolFluid} - T_{Initial}}{T_{CoolFluid} - T_{Final}}\right]$$

For analysis of more complex cooling systems, see the open literature.[1(chp 18)]

After cooling the product in the batch or semi-batch reactor, we pump it from the reactor. The time required to discharge product from the reactor depends upon the volume of product $V_{Product}$ (m^3) and the capacity of the pump, defined as product volumetric flow rate $Q_{Product}$ (m^3/s). The time to discharge product from the reactor is

$$t = \frac{V_{Product}}{Q_{Product}}$$

To clean the batch or semi-batch reactor after discharging product from it, we flush it with solvent, either water or an organic liquid. Flushing a reactor involves multiple time steps, which are

- charge the reactor with flush solvent;
- agitate the solvent inside the reactor to solubilize any product or reactant remaining in the reactor;
- discharge the flush solvent;
- repeat the above steps if necessary.

The time required to charge and discharge solvent are

$$t = \frac{V_{FlushSolvent}}{Q_{FlushSolvent}}$$

where $V_{FlushSolvent}$ is the volume of flush solvent charge to the reactor (m^3) and $Q_{FlushSolvent}$ is the volumetric discharge flow rate of flush solvent (m^3/s). The time required to solubilize any product or reactant remaining in the reactor requires measurement either in a pilot plant reactor or a commercial-sized reactor because scaling such information from laboratory data is difficult.

The batch or semi-batch reactor requires drying after flushing its interior. If our reactor has a jacket, we dry it by applying steam to the jacket and streaming either ambient or hot gas, generally air, through the reactor's interior. If the reactor does not possess a jacket, then we stream hot gas through it.

The drying rate for a batch or semi-batch reactor is

$$R_{Drying} = -\frac{1}{A}\frac{dm}{dt}$$

where R_{Drying} is the drying rate ($kg/m^{2}*s$); A is the interior surface area covered with liquid (m^2); m is the liquid mass (kg); and t is the time (s). When we plot R_{Drying} as a function of time, we see it is a composite function: at early time, R_{Drying} is constant; at later times, R_{Drying} decreases linearly with time. Thus

$$R_{Drying} = R_{ConstantRate} + R_{FallRate}$$

where $R_{ConstantRate}$ occurs when R_{Drying} is constant and $R_{FallRate}$ occurs when R_{Drying} decreases with increasing time. Therefore, the total drying time will be

$$t_T = t_{CR} + t_{FR}$$

where t_T is the total drying time (s); t_{CR} is the drying time during the constant drying rate period (s); and t_{FR} is the drying time during the falling period rate (s).

When R_{Drying} is constant, the liquid inside the reactor is adjusting itself to the drying conditions.[3] During this time period, R_{Drying} is independent of the amount of liquid inside the reactor, implying the presence of a continuous liquid film on the interior surface of the reactor. When this liquid film is no longer continuous, R_{Drying} demonstrates a decreasing rate as time increases. R_{Drying} becomes mass dependent subsequent to this point in time.

We describe the constant drying rate period as

$$R_{\mathrm{ConstantRate}} = -\frac{1}{A}\frac{\mathrm{d}m}{\mathrm{d}t}$$

Rearranging and integrating yields

$$\int_0^{t_{\mathrm{CR}}} \mathrm{d}t = -\frac{1}{AR_{\mathrm{ConstantRate}}}\int_{m_{\mathrm{Initial}}}^{m_{\mathrm{C}}} \mathrm{d}m$$

where t_{C} is the time to reach the point when the liquid film inside the reactor no longer continuously covers the area being dried (s); m_{Initial} is the starting weight of liquid inside the reactor (kg); and m_{C} is the weight of liquid inside the reactor when the liquid film becomes discontinuous (kg). Performing the above integration gives

$$t_{\mathrm{CR}} = -\frac{(m_{\mathrm{C}} - m_{\mathrm{Initial}})}{AR_{\mathrm{ConstantRate}}} = \frac{(m_{\mathrm{Initial}} - m_{\mathrm{C}})}{AR_{\mathrm{ConstantRate}}}$$

Drying rate becomes dependent upon the mass of liquid inside the reactor once the liquid surface becomes discontinuous, at which time R_{Drying} begins decreasing as time increases. Thus

$$R_{\mathrm{FallRate}} = k_{\mathrm{Drying}} m$$

where R_{FallRate} is the drying rate during this time period ($\mathrm{kg/m^{2*}s}$); k_{Drying} is the mass transfer rate constant for drying ($\mathrm{1/m^{2*}s}$); and m is the mass of liquid being dried (kg). But

$$R_{\mathrm{FallRate}} = -\frac{1}{A}\frac{\mathrm{d}m}{\mathrm{d}t}$$

Substituting the penultimate equation into the above equation gives

$$k_{\mathrm{Drying}}\, m = -\frac{1}{A}\frac{\mathrm{d}m}{\mathrm{d}t}$$

Rearranging and integrating yields

$$k_{\mathrm{Drying}} A \int_0^{t_{\mathrm{FR}}} \mathrm{d}t = -\int_{m_{\mathrm{C}}}^{m_{\mathrm{Final}}} \frac{\mathrm{d}m}{m}$$

where t_{FR} is the drying time for the falling rate period (s); m_C is the weight of liquid inside the reactor when the liquid film becomes discontinuous (kg); and m_{Final} is the final weight of liquid inside the reactor. Performing the above integration gives

$$k_{Drying}\,A\,t_{FR} = -\ln\left(\frac{m_{Final}}{m_C}\right) = \ln\left(\frac{m_C}{m_{Final}}\right)$$

Rearranging the last equation yields

$$t_{FR} = \left(\frac{1}{k_{Drying}\,A}\right)\ln\left(\frac{m_C}{m_{Final}}\right)$$

The total drying time is then

$$t_T = \frac{(m_{Initial} - m_C)}{A R_{ConstantRate}} + \left(\frac{1}{k_{Drying}\,A}\right)\ln\left(\frac{m_C}{m_{Final}}\right)$$

Rearranging provides us with

$$t_T = \frac{1}{A R_{ConstantRate}}\left[(m_{Initial} - m_C) + \left(\frac{R_{ConstantRate}}{k_{Drying}}\right)\ln\left(\frac{m_C}{m_{Final}}\right)\right]$$

The final step in a batch or semi-batch reactor cycle is preparing the reactor for use. This step involves inspecting the reactor for cleanliness, reassembling the reactor if piping was decoupled or hatches and manways opened, then pressure testing the reassembled reactor for leaks. We establish this time duration by observing the operating personnel doing it. If the operating personnel undertaking this task are highly trained and skilled at it, then the duration will be minimized. If the operating personnel are new and still learning this task, then the duration will be longer than desired but will decrease as their skill level increases. We identify this time duration as t_{Prep}.

The ultimate time we must record is "down" time. We classify that time when there is no activity occurring at the reactor as down time t_{Down}. Down time occurs during holidays and other periods when the production facility is closed. It also occurs when t_{Prep} ends an hour before shift change during 24/7 operation or ends an hour before quitting time if the production facility staffs only one shift.

We can now calculate the possible number of batches per reactor for a given production facility. If the production facility operates 24/7,

365 days/year, then the number of batches per batch or semi-batch reactor is

$$\text{Batches} = \frac{(365\ \text{days/year})(24\ \text{h/day})}{t_{\text{Charge}} + t_{\text{Heat}} + t_{\text{Reaction}} + t_{\text{Cool}} + t_{\text{Discharge}} + t_{\text{Flush}} + t_{\text{Dry}} + t_{\text{Prep}} + t_{\text{Down}}}$$

where each t has units of h/batch. The above equation determines the capacity of a particular batch or semi-batch reactor at a given production facility.

To optimize the reaction time t_{Rxn} for a batch or semi-batch reactor, we must determine the net profit $(\text{NP})_{\text{Process}}$ for the process or reaction conducted in the reactor. The total operating cost C_{Total} for one batch or semi-batch reactor is

$$C_{\text{Total}} = C_{\text{Charge}} + C_{\text{Heat}} + C_{\text{Rxn}} + C_{\text{Cool}} + C_{\text{Discharge}} + C_{\text{Flush}} + C_{\text{Dry}} + C_{\text{Prep}} + C_{\text{Down}}$$

For a simple, generalized reaction, such as

$$\text{A} + \text{B} \rightarrow \text{P}$$

the cost of reaction is

$$C_{\text{Rxn}} = w_{\text{A}}(A_t - A_{t=0}) + w_{\text{B}}(B_t - B_{t=0})$$

where w_{A} is the unit price per mole of reactant A; w_{B} is the unit price per mole of reactant B; A_t is the moles of reactant A in the reactor at time t; $A_{t=0}$ is the initial number of moles of reactant A in the reactor; and B_t is the moles of reactant B in the reactor at time t; $B_{t=0}$ is the initial number of moles of reactant B in the reactor. The revenue from the product P in the reactor is $w_{\text{P}}P$, where w_{P} is the unit market value per mole for product P. The net profit for the reaction is

$$(\text{NP})_{\text{Rxn}} = w_{\text{P}}P - [w_{\text{A}}(A_t - A_{t=0}) + w_{\text{B}}(B_t - B_{t=0})]$$

Substituting C_{Rxn} into the above equation yields

$$(\text{NP})_{\text{Rxn}} = w_{\text{P}}P - C_{\text{Rxn}}$$

Subtracting all the other costs from $(NP)_{Rxn}$ gives the net profit for the process; thus

$$\begin{aligned}(NP)_{Process} &= (NP)_{Rxn} - C_{Charge} - C_{Heat} - C_{Cool} - C_{Discharge} - C_{Flush} \\ &\quad - C_{Dry} - C_{Prep} - C_{Down} \\ &= w_P P - C_{Rxn} - C_{Charge} - C_{Heat} - C_{Cool} - C_{Discharge} \\ &\quad - C_{Flush} - C_{Dry} - C_{Prep} - C_{Down}\end{aligned}$$

Substituting for C_{Total} yields

$$(NP)_{Process} = w_P P - C_{Total}$$

Differentiating with respect to t_{Rxn} gives

$$\frac{d(NP)_{Process}}{dt_{Rxn}} = w_P \frac{dP}{dt_{Rxn}}$$

Note that we assumed C_{Total} to be independent of t_{Rxn}. Multiplying the above equation by one, i.e., V/V, gives

$$\frac{d(NP)_{Process}}{dt_{Rxn}} = w_P V \frac{d(P/V)}{dt_{Rxn}} = w_P V \frac{d[P]}{dt_{Rxn}}$$

Fitting a polynomial equation to a plot of $[P]$ versus time provides (see Chapter 2)

$$[P] = \alpha + \beta t - \gamma t^2 + \delta t^3 + \cdots$$

which becomes upon differentiation

$$\frac{d[P]}{dt_{Rxn}} = \beta - 2\,\gamma t + 3\,\delta t^2 + \cdots$$

Substituting this last differential equation into the differential equation for $d(NP)_{Process}/dt_{Rxn}$ gives

$$\frac{d(NP)_{Process}}{dt_{Rxn}} = w_P V \frac{d[P]}{dt_{Rxn}} = w_P V(\beta - 2\,\gamma t + 3\,\delta t^2 + \cdots)$$

Setting the above differential equation to zero optimizes $(NP)_{Process}$, namely

$$\frac{d(NP)_{Process}}{dt_{Rxn}} = w_P V(\beta - 2\,\gamma t + 3\,\delta t^2 + \cdots) = 0$$

which gives

$$\beta - 2\,\gamma t + 3\,\delta t^2 + \cdots = 0$$

Solving the above equation for t provides the optimum t_{Rxn} for the process.[4]

PROCESS OPERATING TEMPERATURE

When operating a batch or semi-batch reactor, we must determine

- the best operating temperature;
- the best temperature progression for the reaction

to maximize production, in the case of single reactions, or to maximize yield, in the case of multiple reactions, i.e., in the case where by-products form in the reactor.[5]

For irreversible reactions involving a single product, i.e., for a reaction that does not produce by-products, determining the best operating temperature is relatively simple: it is the highest possible temperature at which reactor materials and solvent, reactant, and product physical properties are compatible.[6] In this case, we choose the highest possible temperature to obtain maximum conversion or to achieve the shortest possible time for reaching a specified reactant conversion.

For exothermic reversible, single reactions, an optimum operating temperature exists because, as the reaction approaches its equilibrium conversion, increasing operating temperature shifts the equilibrium toward higher reactant concentrations.[6] The reverse reaction will be slow at low reactant conversion; however, it increases as reactant conversion, i.e., product formation, increases. In addition, the activation energy of the reverse reaction, E_R, is greater than the activation energy of the forward reaction, E_F; otherwise, an economic amount of product does not accumulate in the reactor. Because $E_R > E_F$, the rate of the reverse reaction increases faster with increasing temperature than does the rate of the forward reaction. Therefore, we should use a temperature sequence when operating a reversible, exothermic reaction. Initially, we use a high temperature to produce product at an economically acceptable rate, but, as conversion increases, we decrease the operating temperature to reduce the reverse reaction rate relative to the forward reaction rate, which preserves the product already formed in the reactor.[6(p380),7]

We can quantify our reasoning as follows: consider the reversible reaction

$$A \Leftrightarrow B$$

For a batch reactor, the component balance for reactant A, in terms of moles, is

$$-\frac{dn_A}{dt} = R_A$$

where R_A is

$$R_A = k_F n_A - k_R n_B$$

with k_F being the forward rate constant (1/s); n_A, the moles of reactant A; k_R, the reverse rate constant (1/s); and n_B, the moles of product B. At any given time

$$n_B = n_{B/t=0} + n_{A/t=0} - n_A$$

where $n_{A/t=0}$ and $n_{B/t=0}$ are the initial moles of reactant A and product B in the batch reactor. Of course, at $t = 0$, $n_{B/t=0}$ is zero; therefore, the above equation becomes

$$n_B = n_{A/t=0} - n_A$$

Substituting this last equation into the equation for R_A gives

$$R_A = k_F n_A - k_R(n_{A/t=0} - n_A)$$

Defining conversion as

$$x = \frac{n_{A/t=0} - n_A}{n_{A/t=0}} = \frac{n_B}{n_{A/t=0}}$$

then rearranging that definition in terms of n_A and n_B gives

$$\begin{aligned} n_B &= x n_{A/t=0} \\ n_A &= (1 - x) n_{A/t=0} \end{aligned}$$

Substituting the above definitions into the last equation for R_A, then rearranging yields

$$R_A = n_{A/t=0}[(1 - x)k_F - x k_R]$$

Differentiating R_A with respect to operating temperature T at a specified conversion x, then setting the differential equal to zero to determine the optimum operating temperature, yields

$$\frac{dR_A}{dT} = n_{A/t=0}\frac{d}{dT}[(1-x)k_F - xk_R] = 0$$

or

$$\frac{d}{dT}[(1-x)k_F - xk_R] = 0$$

The Arrhenius equation relates the rate constants to operating temperature as

$$k_F = A_F e^{-(E_F/R_g T)}$$

and

$$k_R = A_R e^{-(E_R/R_g T)}$$

where the subscript F references the forward reaction and the subscript R references the reverse reaction; A is the "frequency factor" or "collision parameter" for each respective reaction; E is the "activation energy" for each respective reaction; R_g is the gas constant; and T is the operating temperature. Differentiating each rate constant with respect to T yields

$$\frac{dk_F}{dT} = -A_F\left(\frac{E_F}{R_g}\right)e^{-(E_F/R_g T)}$$

$$\frac{dk_R}{dT} = -A_R\left(\frac{E_R}{R_g}\right)e^{-(E_R/R_g T)}$$

Substituting the above differential equations into the optimization equation yields

$$(1-x)\left[-A_F\left(\frac{E_F}{R_g}\right)e^{-(E_F/R_g T)}\right] - x\left[-A_R\left(\frac{E_R}{R_g}\right)e^{-(E_R/R_g T)}\right] = 0$$

Rearranging the above equation, then simplifying gives

$$\frac{x}{1-x} = \left(\frac{A_F E_F}{A_R E_R}\right)e^{(E_R - E_F)/R_g T}$$

Solving this last equation for T yields the optimum operating temperature for a specified conversion x; thus

$$T_{\text{Optimum}} = \frac{E_R - E_F}{R_g \ln[(A_R E_R/A_F E_F)(x/(1-x))]}$$

This equation yields the optimum temperature at each conversion for a reversible, exothermic reaction. Note that we generally conduct reversible, exothermic reactions in tubular reactors since they respond faster to a desired temperature change than do batch reactors. However, such reactions can be performed in batch reactors equipped with external heat exchangers for removing heat. The circulation flow rate must be high to achieve the desired operating temperature profile.

For a reversible, endothermic reaction, $E_F > E_R$; thus, the forward reaction rate increases relative to the reverse reaction rate at any operating temperature, independent of conversion. Therefore, the optimum temperature sequence is the maximum allowable operating temperature for the process.[7]

For most multiple reaction processes in which by-products form, mathematical analysis can be daunting and requires numerical solution. However, we can use deductive reasoning to obtain insight into the optimum temperature profile required for an economically viable process.

Consider the parallel reaction scheme

$$X + Y \rightarrow P$$
$$X + Y \rightarrow BP$$

where P identifies desired product and BP identifies undesired by-product. We can optimize this reaction scheme for product formation or for product yield. If the energy of activation for by-product BP (E_{BP}) is greater than the energy of activation for product P (E_P), then a low operating temperature favors P formation. However, a low operating temperature extends the time required to complete the reaction, thereby necessitating a large reactor. Thus, to maximize product formation, we would use a large reactor and operate the process at a constant low temperature. To maximize product yield, we utilize a low operating temperature initially, then increase the operating temperature as the concentrations of X and Y decrease.

Consider the consecutive reaction scheme

$$X + Y \rightarrow P \rightarrow BP$$

where the symbols have been defined above. If $E_{BP} > E_P$, then a low, continuous operating temperature maximizes product yield, but at a slow rate.[5(p129)] To maximize product formation rate, specifying a high initial operating temperature permits the rapid accumulation of P; to conserve that accumulated P, we decrease the operating temperature while continuing to consume X and Y, which requires a reactor with an external cooler capable of a high circulation rate.[6(p382),8] If $E_P > E_{BP}$, then a high, continuous operating temperature maximizes product yield.[5(p129)] The chemical engineering literature abounds with many other such examples.[9,10]

SUMMARY

This chapter discussed process cycle time, which is the total time from charging feed to a batch or semi-batch reactor to the next feed charge of the same reactor, and its optimization. This chapter also discussed the reaction time required to optimize the profitability of a given batch or semi-batch reactor. It finished by presenting scenarios for maximizing product formation and quality through process temperature optimization.

REFERENCES

1. Kern D. *Process heat transfer*. New York, NY: McGraw-Hill Book Company; 1950. p. 626–627.
2. Coker A. *Modeling of chemical kinetics and reactor design*. Boston, MA: Gulf Professional Publishing; 2001. p. 637–639.
3. McCabe W, Smith J. *Unit operations of chemical engineering*. 3rd ed. McGraw-Hill Book Company; 1976. p. 785.
4. Aris R. *The optimal design of chemical reactors*. New York, NY: Academic Press; 1961.
5. Denbigh K. *Chemical reactor theory: an introduction*. Cambridge, UK: Cambridge University Press; 1966. p. 126.
6. Froment G, Bischoff K. *Chemical reactor analysis and design*. New York, NY: John Wiley & Sons, Inc.; 1979. p. 377.
7. Smith J. *Chemical engineering kinetics*. 3rd ed. New York, NY: McGraw-Hill Book Company; 1983. p. 259.
8. Rase H. *Chemical reactor design for process plants: volume 1—principles and techniques*. New York, NY: John Wiley & Sons, Inc.; 1977. p. 567–569.
9. Fournier C, Groves F. Isothermal temperatures for reversible reactions. *Chem Eng* 1970;**77** (3):121–5.
10. Fournier C, Groves F. Rapid method for calculating reactor temperature profiles. *Chem Eng* 1970;**77**(13):157–9.

Printed in the United States
By Bookmasters